D0559516

Control Systems for Heating, Ventilating and Air Conditioning

Control Systems for Heating, Ventilating and Air Conditioning

Second Edition

Roger W. Haines

Project Engineer
Ziel-Blossom and Associates
Cincinnati, Ohio

VAN NOSTRAND REINHOLD ENVIRONMENTAL ENGINEERING SERIES

VAN NOSTRAND REINHOLD COMPANY
NEW YORK   CINCINNATI   ATLANTA   DALLAS   SAN FRANCISCO
LONDON   TORONTO   MELBOURNE

Van Nostrand Reinhold Company Regional Offices:
New York   Cincinnati   Atlanta   Dallas   San Francisco

Van Nostrand Reinhold Company International Offices:
London   Toronto   Melbourne

Copyright © 1977 by Litton Educational Publishing, Inc.

Library of Congress Catalog Card Number: 76-58531
ISBN: 0-442-23031-1

All rights reserved. Certain portions of this work copyright © 1971
by Litton Educational Publishing, Inc.
No part of this work covered by the copyright hereon may
be reproduced or used in any form or by any means—graphic, electronic, or
mechanical, including photocopying, recording, taping, or information storage
and retrieval systems—without permission of the publisher.

Manufactured in the United States of America

Published by Van Nostrand Reinhold Company
450 West 33rd Street, New York, N.Y. 10001

Published simultaneously in Canada by Van Nostrand Reinhold Ltd.

15 14 13 12 11 10 9 8 7 6 5 4 3 2

Library of Congress Cataloging in Publication Data

Haines, Roger W
   Control systems for heating, ventilating, and air
conditioning.

   (Van Nostrand Reinhold environmental engineering
series)
   Bibliography: p.
   Includes index.
   1. Heating—Control.   2. Air conditioning—Control.
I. Title.
TH7466.5.H34   1977   697'.07   76-58531
ISBN 0-442-23031-1

Van Nostrand Reinhold Environmental Engineering Series

ADVANCED WASTEWATER TREATMENT, by Russell L. Culp and Gordon L. Culp

ARCHITECTURAL INTERIOR SYSTEMS–Lighting, Air Conditioning, Acoustics, John E. Flynn and Arthur W. Segil

SOLID WASTE MANAGEMENT, by D. Joseph Hagerty, Joseph L. Pavoni and John E. Heer, Jr.

THERMAL INSULATION, by John F. Malloy

AIR POLLUTION AND INDUSTRY, edited by Richard D. Ross

INDUSTRIAL WASTE DISPOSAL, edited by Richard D. Ross

MICROBIAL CONTAMINATION CONTROL FACILITIES, by Robert S. Runkle and G. Briggs Phillips

SOUND, NOISE, AND VIBRATION CONTROL, by Lyle F. Yerges

NEW CONCEPTS IN WATER PURIFICATION, by Gordon L. Culp and Russell L. Culp

HANDBOOK OF SOLID WASTE DISPOSAL: MATERIALS AND ENERGY RE-COVERY, by Joseph L. Pavoni, John E. Heer, Jr., and D. Joseph Hagerty

ENVIRONMENTAL ASSESSMENTS AND STATEMENTS, by John E. Heer, Jr. and D. Joseph Hagerty

ENVIRONMENTAL IMPACT ANALYSIS: A New Dimension in Decision Making, by R. K. Jain, L. V. Urban and G. S. Stacey

CONTROL SYSTEMS FOR HEATING, VENTILATING, AND AIR CONDITIONING (Second Edition), by Roger W. Haines

v

Van Nostrand Reinhold Environmental Engineering Series

THE VAN NOSTRAND REINHOLD ENVIRONMENTAL ENGINEERING SERIES is dedicated to the presentation of current and vital information relative to the engineering aspects of controlling man's physical environment. Systems and subsystems available to exercise control of both the indoor and outdoor environment continue to become more sophisticated and to involve a number of engineering disciplines. The aim of the series is to provide books which, though often concerned with the life cycle—design, installation, and operation and maintenance—of a specific system or subsystem, are complementary when viewed in their relationship to the total environment.

Books in the Van Nostrand Reinhold Environmental Engineering Series include ones concerned with the engineering of mechanical systems designed (1) to control the environment within structures, including those in which manufacturing processes are carried out, (2) to control the exterior environment through control of waste products expelled by inhabitants of structures and from manufacturing processes. The series will include books on heating, air conditioning and ventilation, control of air and water pollution, control of the acoustic environment, sanitary engineering and waste disposal, illumination, and piping systems for transporting media of all kinds.

Preface to the Second Edition

In the five and one-half years which have elapsed since the publication of this book we have seen two developments of great importance to the control industry. The first was the improvement and simplification of the computerized supervisory system—a development we could foresee in 1971 though we were unable to describe the shape it would take. The present situation in this field is described in Chapter 10 which has been completely rewritten and expanded.

The second important event was the "energy crisis"—an event which made many more people conscious of the importance of A/C systems and controls in energy consumption. I considered adding a section to this edition on controls for energy conservation, but on rereading the book, I discovered that there are already many allusions to economy operation and energy use, which simply shows that, as engineers, we have been aware of the problem for some time, but failed to get the message across to the right people. I'm not convinced that it is being heard yet, but there is a great deal of information in this book on energy conservation.

The success of the first edition has been a source of great personal pleasure and I hope that this revised edition will be even more helpful than the first.

Roger W. Haines

Preface to the First Edition

This book is intended for the guidance of the engineer who is designing a heating, ventilating or air conditioning system and wants a simple, practical explanation of how best to control that system. It could also be used as a supplemental text in a college or technical school course on refrigeration and air conditioning.

It does not include mathematical analyses of control systems, response factors, Fourier transforms and the like. These are adequately and thoroughly covered in a number of up-to-date college-level texts.

What is presented here is an elementary but comprehensive explanation of control system theory, control hardware, and both simple and complex control systems. There are also discussions on supervisory controls and the use of computers in control systems.

Throughout, the reader should be aware of the interrelationship between the HVAC controls, the HVAC system, the electrical power system and the building. There have been examples of control systems which failed to control because inherent deficiencies in the building or HVAC system made it uncontrollable; or of control failures because the designer failed to make his intent clear to others. It is hoped that the reader will be helped to avoid some of these pitfalls.

Roger W. Haines

January, 1971

Acknowledgments

Nothing gets done without help and I've had some
of the best. First and foremost, Frank Bridgers and
Don Paxton provided my basic training in consulting
engineering and control design. My debt to them
is large. Many friends in ASHRAE, particularly those
on TC 10.1 (Controls) gave encouragement and help.
Tom Guiterrez, a colleague at Collins Radio, served
as sounding board, critic, research assistant and
advisor. Don Bahnfleth, editorial director of
Heating, Piping and Air Conditioning magazine read
the manuscript and gave much good advice. My
wife encouraged and helped me as she has so often,
besides typing a good part of the manuscript.

Professor Clark Pennington and Dr. Joseph
Gartner read and criticized the manuscript and their
comments were most helpful.

Several manufacturers graciously gave permission
to reproduce parts of their literature. These include
Barber-Coleman Co., Honeywell Inc., Johnson
Service Co., and Penn Controls Inc.

Roger W. Haines

Contents

Control Systems for Heating, Ventilating and Air Conditioning

1 Control Theory and Terminology

1.1 INTRODUCTION

The purpose of this book is to discuss the design of control systems for heating, ventilating and air conditioning systems. Its intent is to help you, the reader, develop an understanding of controls and control systems, and the air conditioning systems to which they are applied. Out of this you should develop a philosophy of design which will enable you to cope not only with the basic systems discussed here, but with the unusual and special requirements which continue to arise as air conditioning becomes more sophisticated.

The term "heating, ventilating and air conditioning" (HVAC) covers a wide range of equipment, from, for example, a kerosene stove, to the large and sophisticated complex of equipment required for the new World Trade Center in New York City.

"Control" likewise may vary from the handwheel adjustment of the kerosene stove wick to the elaborate, computerized system in the World Trade Center.

This book will discuss as many as possible of the various HVAC systems in use today, together with the methods of control which may be applied to them. The word "together" should be emphasized. The HVAC system, its control system and the building in which they are installed are inseparable parts of a whole. They interact with

1

one another in many ways, so that neglect of any element may cause a partial or complete loss of controllability.

Unfortunately, there have at times been examples of this neglect, with resultant regret on the part of all concerned. It is possible to design good HVAC systems and controls at a reasonable cost. No HVAC system is better than its controls and the building in which it is installed. As designers we have a duty to provide the owner with the best possible system within budget limits—not necessarily the cheapest. The cheapest may be the most expensive in the long run, in operating cost as well as owner dissatisfaction. The "best" system is one that will provide the required degree of comfort for the application. That degree of comfort is, of course, a function of the application; that is, we should expect closer control of temperature in a hotel bedroom than in the same hotel's kitchen.

1.2   WHAT IS "CONTROL"?

While many HVAC control systems appear to be, and are, complicated, the most elaborate system may be reduced to a few fundamental elements. Let's get back to that kerosene stove. We are cold, so we strike a match and light the wick (after checking the fuel supply). We turn the wick up high. After a while it gets warm. We sense this and turn the handwheel to lower the wick and get less heat. Or, we may turn it clear down, shutting off the heat entirely.

Here are all the elements of a closed-loop control system. The "controlled variable" is the air temperature in the room. The "process plant" is the stove. The wick is the "controlled device." The sensor and controller are represented by the person in the room. You will note that a man is not a really sensitive controller. Nonetheless he performs the basic function of the sensor-controller, which is to measure the controlled variable, compare it with a "set point" (here, the personal sensation of comfort) and adjust the controlled device.

Notice that only three elements are necessary for a control system: sensor, controller and controlled device. These need only a process plant to give them meaning. In this text the definition of the process plant is limited to heating, ventilating and air conditioning systems.

1.3   ELEMENTARY CONTROL SYSTEM

Figure 1–1 illustrates an elementary control system. This shows air flowing through a heating coil in a duct. The sensor measures

Figure 1-1 Elementary control system.

the temperature of the air downstream of the coil and passes the information to the controller. The controller compares the air temperature with a set point and sends a signal to open or close the hot-water valve (the controlled device) as required to maintain a correspondence between the air temperature and the set point. This is a "closed-loop" system, in which the change in temperature caused by a change in valve position (and/or load) will be sensed and additional adjustments will be made as necessary. The air temperature is known as the "controlled variable."

Most control systems fall in the "closed-loop" classification but "open-loop" systems are sometimes used. In an open-loop system the sensor is not directly affected by the action of the controlled device. A homely example of such a system is the electric blanket, where the thermostat senses room and not blanket temperature.

Keep in mind that, regardless of apparent complexity, all control systems may be reduced to these essential elements. Most complications occur in an attempt to obtain "better" control; that is, to maintain the controlled variable as close to the set point as possible. One of the cardinal rules of control system design is to "keep it simple," and avoid piling relays on resets on multiple sensors.

1.4 PURPOSES OF CONTROL

It is usually thought that the purpose of the automatic control system is to provide control of the temperature and/or humidity in a space. But these are not the only functions which the system can serve:

It can also control the relative pressure between two spaces, a very useful attribute in preventing the spread of contamination.
Safety controls prevent operating of equipment when in an unsafe

condition. They can also trigger visual or audible alarms to alert operating personnel to those conditions.

The HVAC system operates most economically when equipment capacity is closely matched to load, and this can be better accomplished by an automatic control system than manually.

A completely automatic system with changeover controls, interlocks, and internal monitoring and compensating controls minimizes human intervention and, therefore, the chance of human error.

It can be seen, then, that the purposes of control systems are many and varied. Now, how are the required functions carried out?

1.5   CONTROL ACTION

To satisfy the need for various kinds of control response, there are available a number of types of control actions. These may be broadly classified as follows.

1.5.1   Two-Position or On-Off Action (Figure 1–2)

This is the simplest and most obvious. One example of two-position action is a relay which is open or closed, with no intermediate position. An example of on-off action is a thermostat starting and stopping a ventilating fan motor. Any two-position controller needs a *differential* to prevent "hunting," or too-rapid cycling. This differential is the difference between the setting at which the controller operates to one position and the setting at which it changes to the other. In a thermostat this is expressed in degrees of temperature. The *differential* setting of any controller is usually somewhat less than the *operating differential* of the HVAC system due to the lag of the instrument and the system.

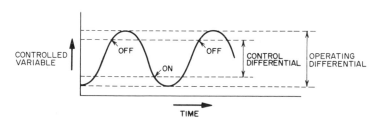

Figure 1-2   Two-position control (low limit).

Figure 1-3 Floating action.

1.5.2 Timed Two-Position Action

This is used to reduce the operating differential by artificially shortening "on" or "off" time in anticipation of system response. A heating thermostat may be provided with a small internal heater which is energized during "on" periods, thereby giving a false signal to the thermostat. This is called "heat anticipation."

1.5.3 Floating Action (Figure 1–3)

This term refers to a controlled device which can stop at any point in its stroke and can be reversed without completing its stroke. The controller must have a "dead spot" or neutral zone in which it sends no signal but allows the device to "float" in a partly open position. For good operation this system requires a rapid response in the controlled variable; otherwise it will not stop at an intermediate point.

1.5.4 Proportional Action (Figure 1–4)

Proportional action is simply floating action with *feedback* added. Feedback means that the controlled device actuator moves only enough to satisfy the change in the controlled variable. It is not necessary to wait for the controlled variable to respond. The term "modulating control" usually refers to proportional control, though strictly speaking, floating control is also "modulating."

With proportional control we meet some new terms:

Throttling range is the amount of change in the controlled variable required to run the actuator of the controlled device from one end of its stroke to the other.

Figure 1-4   Proportional action.

Set point is the controller setting and is the desired value of the controlled variable.

Control point is the actual value of the controlled variable. If the control point lies within the throttling range of the controller it is said to be in control. When it exceeds the throttling range it is said to be out of control.

Offset or *control point shift* is the difference between the set point and the control point. This is sometimes called drift, droop, or deviation. The amount of offset theoretically possible is determined by the throttling range, but this value may be exceeded in out-of-control situations.

1.5.5   Proportional Action with Automatic Reset (Figure 1–5)

This simply means that the control point is automatically returned to the original set point whenever any offset occurs. The *effect* of automatic reset is to increase the magnitude of the change in the controlled variable as the rate of change of the control point increases. Thus the return of the control point to the original set point is accelerated. This provides more accurate control of the controlled variable, with the stability provided by wide throttling range.

Figure 1-5   Proportional action with reset.

Reset rate is the number of times per unit of time the reset action occurs. Reset rate is usually adjustable and to avoid unstable operation must be carefully adjusted to the controlled system characteristics. That is, if the controlled system has quick response the reset rate must be faster.

Stability refers to the ability of the system to maintain the controlled variable at or near the set point without requiring the controller to "hunt," that is, change continuously from one extreme to the other. Thus it becomes a function of the entire process plant and environment as well as the control system. What this means to us is: If the HVAC system and building have a slow response to changes in the controlled variable, then a wide differential or throttling range is necessary to provide stability. If system and building response are rapid, then a smaller differential may be used.

1.6 ENERGY SOURCES FOR CONTROL SYSTEMS

The above-described control actions may be accomplished by various means. Control systems may be electric, electronic, pneumatic, fluidic, hydraulic, self-contained or combinations of these. Chapters 2, 3, 4 and 5 will discuss the control elements in some detail. A brief description here will serve to introduce the next chapters.

1.6.1 Electric Systems

Electric systems provide control by starting and stopping the flow of electricity or varying the voltage and current by means of rheostat or bridge circuits.

1.6.2 Electronic Systems

These systems use very low voltages (15 V or less) and currents for sensing and transmission, with amplification by electronic circuits or servo-mechanisms as required for operation of controlled devices.

1.6.3 Pneumatic Systems

Pneumatic systems usually use low-pressure compressed air. Changes in output pressure from the controller will cause a corresponding position change at the controlled device.

1.6.4   Hydraulic Systems

These are similar in principle to pneumatic systems but use a liquid or gas rather than air. These systems are usually closed, while pneumatic systems are open (some air is wasted).

1.6.5   Fluidic Systems

Fluidic systems use air or gas and are similar in operating principles to electronic as well as pneumatic systems.

1.6.6   Self-Contained System

This type of system incorporates sensor, controller and controlled device in a single package. No external power or other connection is required. Energy needed to adjust the controlled device is provided by the reaction of the sensor with the controlled variable.

1.7   MEASUREMENT

You will have noted that all the control actions described above depend first on *measurement* of a controlled variable. This matter of accurate and rapid measurement is the most serious problem in the control industry. While it is essential for proper control, it is very difficult to obtain an accurate and instantaneous reading, especially if the property being measured is fluctuating or changing very rapidly.

It is, therefore, necessary to analyze very carefully what is actually being measured, how it may vary with time in operation and how great a degree of accuracy is necessary in the measurement.

Thermostats will be affected by the presence or absence of air motion (drafts), the temperature of the surfaces on which they are mounted (if greatly different from the air temperature), the mass of the sensing element and the presence of radiant effects from windows or hot surfaces. For a residence or an office a variation of one or two degrees on either side of the set point may be acceptable. For a standards laboratory (Chapter 9) a variation of $\pm 0.5°F$ may be unacceptable. And, of course, while the thermostat in a standards lab may show a variation of only $0.2°$ to $0.1°$, across the room, 20 ft way, the variation may very well be $0.5°$ or even $1°$.

A pressure sensor which is located at a point of turbulence (such as a turn or change of pipe size) in the fluid can never provide accurate or consistent readings. For this purpose a long straight run of duct or pipe is generally required. Straightening vanes can be used where long straightaways are not possible.

Another difficulty which may arise in measurement or control is delay due to the distance over which the signal must be transmitted. Pneumatic signals will travel only at sonic speeds and are subject to fluid friction losses. Electric signals may become seriously attenuated by resistance in long lines.

Several books have been written on sensors and measuring problems. (See, for example, references 2 and 8 in the Bibliography.) It is not the purpose of this book to go into such detail; but we hope you will study carefully the location of all sensors and their relationship to the rest of the system.

1.8 SUMMARY

This chapter has discussed the elements of a control system, the basic types of control action, and the energy sources commonly used for controls. Many control systems use combinations of these energy types. Interface between unlike types of energy is provided by relays or transducers. The next chapters discuss the various types of control elements in detail. The chapters following the element descriptions will consider how these may be combined to make control *systems* of varying degrees of complexity to perform specific functions.

2 Pneumatic Control Devices

2.1 INTRODUCTION

Chapter 1 dealt with the fundamentals of control circuits. This chapter and the three following will consider various types of control devices: pneumatic, electric, electronic and fluidic. Succeeding chapters will discuss the use of these devices in control systems.

No attempt will be made here to provide an exhaustive catalog of control instruments. Rather, some of the basic principles of operation and general classifications available will be considered. With this as background, it will then be possible for the reader to evaluate a catalog description of a control device with a reasonable degree of understanding.

For simplicity, the chapters which follow classify control devices by energy type: pneumatic, electric, electronic and fluidic. This is followed by a discussion of flow control devices (valves and dampers) since these are essentially independent of the actuator.

Schematic diagrams are used, which avoid details that may obscure the situation and are peculiar to one manufacturer. These details are readily available from the manufacturer's service manuals.

2.2 PNEUMATIC CONTROL DEVICES

Pneumatic controls are powered by compressed air, usually 15 to 20 psig pressure, although higher pressures are occasionally used for operating very large valves or dampers. Pneumatic devices are inherently modulating, since air pressure may easily and simply be provided with infinite variation over the control range. Because of their simplicity and low cost, pneumatic controls are frequently found on commercial and industrial installations using more than 8 or 10 devices.

Pneumatic devices available include sensors, controllers, actuators, relays and transducers. These are described in some detail in the paragraphs that follow. The principles described form the basis of most manufacturers' designs, although, of course, each has variations.

2.2.1 Definitions

There are a few definitions which are peculiar to pneumatic devices. Some of these follow:

1. Direct-acting: A controller is direct-acting when an increase in the control signal (temperature, pressure, etc.) results in an increase in the output air pressure.
2. Reverse-acting is the opposite of direct-acting, that is, an increase in the control signal results in a decrease in output air pressure.
3. SCFM: Standard cubic feet per minute. This refers to air at standard atmospheric pressure of 14.7 psia and a temperature of 70°F. For ease of comparison most air compressors are rated in SCFM.
4. Psia: Pounds per square inch, absolute pressure.
5. Psig: Pounds per square inch, gauge. At standard atmosphere, psig + 14.7 = psia.
6. SCIM: Standard cubic inches per minute. Similar to SCFM, but SCIM is usually used to describe pneumatic device air consumption. One SCFM equals 1728(12³) SCIM.

Other terms such as bleed, non-bleed, submaster, are more easily described along with the device description.

2.2.2 Bleed-Type Controllers

The bleed-type controller (Figure 2–1) is the simplest type of pneumatic control device. The control air output pressure is a

Figure 2-1   Bleed-type controller.

function of the amount of air flowing through the nozzle; as the nozzle flow is restricted the output pressure rises. Nozzle flow is controlled by varying the position of a flapper valve, in accordance with the sensor requirement. The device may be direct- or reverse-acting, depending on the sensor linkage, and is inherently proportional. For two-position action a relay is required. The output signal may go directly to an actuator, or may be used as input to a relay-controller.

2.2.3   Relay-Type Controllers

Relay-type controllers may be either non-bleed or pilot-operated bleed-type. The non-bleed controller uses air only when exausting the line to the controlled device.

Figure 2–2 is a schematic diagram of a non-bleed controller. A positive movement from the sensor (increase in temperature or pressure) will cause an inward movement of the diaphragm and a downward movement on the right end of the lever, which raises the

Figure 2-2   Non-bleed controller.

Figure 2-3   Proportional relay controller, pilot-bleed type. *(Courtesy Johnson Service Company.)*

other end of the lever and allows the supply air valve to open. This increases the output pressure and the pressure in the valve chamber. As the chamber pressure increases it acts on the diaphragm to offset the pressure from the sensor (negative feedback), so that when equilibrium is attained the supply valve closes and the balanced pressure becomes the output to the controlled device. This is, therefore, a proportional direct-acting controller with negative feedback. A further positive sensor movement will cause a rebalancing at some higher output pressure. A negative sensor movement will decrease the pressure on the right end of the lever, allowing the exhaust valve to open until the chamber and sensor pressures are again in balance. A simple reversal of the linkage will change the controller to reverse-acting.

The bleed-type piloted controller utilizes a reduced-airflow bleed-type pilot circuit combined with an amplifying non-bleed relay to produce a sensitive, fast-acting control device. The controller can be adjusted to produce a large change in output for a small change in pilot pressure, and can be provided with negative feedback for proportional action or positive feedback for two-position action.

The proportional arrangement is as shown in Figure 2–3. The orifice plate is provided to restrict the flow of air to the pilot chamber. The control port may be partially or completely restricted by the flapper valve, which is operated by the sensor.

In Figure 2–4 the various operating conditions are shown: In Figure 2–4(A) the control port is open, the pilot chamber pressure is essentially zero, the exhaust port is open and output is zero. In Figure 2–4(B) the flapper valve has been partially closed and the pressure

Figure 2-4 Operation of proportional relay. *(Courtesy Johnson Service Company.)* (A) When the control port is open, the exhaust valve between the control and exhaust chambers is open. Thus, air in the control chamber is at zero gage or atmospheric pressure. The supply valve is held closed by a spring and supply pressure. (B) When the sensing element moves closer to the control port, pressure begins to increase in the pilot chamber. At 3 psig, pilot pressure overcomes the force of the opposing spring and closes the exhaust seat.

begins building up. At some initial pressure, usually 3 psig, the spring allows the exhaust seat to close. In Figure 2–4(C), the pressure continues to increase, pushing the pilot diaphragm down and opening the supply valve. Air flows into the control chamber and to the output line. As the pressure in the control chamber increases the pilot pressure is opposed (Figure 2–4(D)) (negative feedback) until the pressures balance and the supply valve closes.

The two-position controller (Figure 2–5) is similar, but with some important differences in valve and spring arrangement. Figure 2–6 shows the operation: In Figure 2–6(A) the control port is open, and spring pressure holds the supply port open and exhaust port closed. This is a reverse-acting controller. As the control port is partially

(C)

(D)

Figure 2-4 (Continued) (C) As pilot pressure continues to increase, it forces the pilot diaphragm down and opens the supply port. This allows supply air to flow to the control line. (D) Pressure now increases in the control chamber, and acts against the control diaphragm to oppose pilot pressure. When the total of forces in each direction is equal, the supply valve is closed and the controller is balanced.

closed, pilot pressure increases, forcing the pilot diaphragm down and closing the supply valve (Figure–6(B)). A further increase in pilot pressure opens the exhaust valve, causing a decrease in pressure in the control chamber. This decreases the force opposing the pilot pressure (positive feedback) and allows the exhaust valve to open completely and remain open until pilot pressure is reduced (Figure 2–6(C)).

Typical controllers of these types are provided with a broad sensitivity adjustment. For example, a temperature controller may have an output change of as little as 1 psi per degree of temperature, or as great as 10 psi per degree.

The mounting of the sensing element determines whether the controller is direct- or reverse-acting. Some devices may be changed in the field; others are fixed and cannot be changed.

Figure 2-5   Two-position relay controller, pilot-bleed type. *(Courtesy Johnson Service Company.)*

2.2.4   Sensing Elements

Sensors have been referred to only in passing up to this point, noting that the controllers can be operated in response to various stimuli, such as temperature, pressure or humidity. A brief discussion of the more common sensing elements is in order.

2.2.4.1   Temperature-Sensing Elements include bimetal; vapor-filled bellows; and liquid-, gas- or refrigerant-filled bulb and capillary.

The bimetal element is the simplest and most often encountered sensor. It consists of two thin strips of dissimilar metals fused together. Since the two metals expand at different rates, the element bends as the temperature changes. The resulting motion can typically be used to vary the pressure at a pneumatic control port or to open and close an electrical contact. The device is most often made of brass (with a high coefficient of expansion) and invar metal which has a very low expansion coefficient. Figure 2–7 shows the movement of a bimetal element as it is heated. The bimetal may also be shaped in a spiral coil to provide a rotary motion output. The bimetal is often used in wall-mounted thermostats where it will sense ambient air temperature.

Figure 2-6   Operation of a two-position relay. *(Courtesy Johnson Service Company.)*
(A) When the control port is open, supply air flows through the supply chamber and
out the control chamber and the controlled devices. Spring and air pressure combine
to hold the supply ball valve open and the exhaust valve closed. (B) As the control
port is closed by the element, pressure in the pilot chamber increases. When pilot
pressure is greater than the opposing forces, the pilot diaphragm moves, seating the
supply ball valve. (C) Further movement of the pilot diaphragm opens the exhaust
ball valve, so that control air is exhausted. This action reduces the forces opposing
the pilot pressure (positive feedback) and causes the exhaust valve to open fully and
remain open until pilot pressure is reduced (by opening the control port). The
sequence is then reversed.

Figure 2-7 Bimetal sensor.

Another ambient temperature sensor is the vapor-filled bellows. The bellows is usually made of brass and filled with a thermally sensitive vapor, which will not condense at the temperatures encountered. Temperature changes will cause the bellows to expand and contract. An adjustable spring is used to control set point and limit expansion. The resulting movement can be used directly or through a linkage.

Bulb and capillary elements are used where temperatures must be measured in ducts, pipes, tanks or similar locations remote from the controller. There are three essential parts of this device: bulb, capillary and diaphragm operating head. The fill may be a liquid, gas or refrigerant, depending on the temperature range desired. Expansion of the fluid in the heated bulb exerts a pressure which is transmitted by the capillary to the diaphragm and there translated into movement. (Figure 2–8.)

The sensing bulb may be only a few inches long, as used in a pipe or tank, or it may be as long as 20 ft when used to sense average air temperature in a duct of large cross section. Special long bulbs are used for freeze protection. These will give a reaction if any one-foot section is exposed to freezing temperatures. Refrigerant is used in this type of bulb, since the refrigerant will condense at freezing temperatures, causing a sharp decrease in pressure.

Temperature-compensated capillary tubes are used to avoid side effects from the ambient temperature around the capillary. Capillaries may be as long as 25 or 30 ft.

Figure 2-8 Bulb and capillary sensor.

Pressure on the diaphragm operates against an adjustable spring to move a post and/or lever. (See also 3.1.1.1 electric elements.)

2.2.4.2 Pressure-Sensing Elements include diaphragms, bellows and bourdon tubes.

The diaphragm is a flexible plate, sealed in a container so that fluid cannot leak past it. A force applied to one side will cause it to move or flex. A spring usually operates to control the movement and return the diaphragm to its initial position when the force is removed. Some diaphragm materials will spring back to the original shape without help. A wide variety of materials are used, to cope with the various temperatures, pressures and fluids encountered.

A bellows is a diaphragm which is joined to the container by a series of convolutions (folds) so that a greater degree of movement may be obtained (Figure 2–9). The bellows may be completely sealed, as in a temperature-sensitive unit, or it may have a connection for sensing pressure, either internally or externally. The bellows acts like a spring, returning to its original shape when the external force is removed. Frequently a separate spring is added, for adjustment and to increase reaction speed.

The bourdon tube (Figure 2–10) is widely used in pressure gages and other pressure instruments. It consists of a flattened tube bent into circular or spiral form. One end is connected to the pressure source and the other end is free to move. As pressure is increased the tube tends to straighten out, and this movement may be used, through an appropriate linkage, to position an indicator or actuate a controller.

2.2.4.3. Humidity-Sensing Elements are made of hygroscopic materials, which change size in response to changes in humidity. An element similar to a bimetal is made of two strips of unlike woods glued together. The different rates of hygroscopic expansion will cause the strip to bend as humidity changes. Yew and cedar woods are frequently used for this purpose.

Elements made of animal membrane, special cloths or human hair

Figure 2-9 Bellows sensor.

Figure 2-10 Pressure sensor, bourdon-tube type. *(Courtesy Johnson Service Company.)*

will increase or shorten their length as humidity changes, with the resulting movement mechanically amplified. Electronic humidity sensors are made of moisture-sensitive compounds which change their electrical conductivity with a change in humidity. (See paragraph 3.2.2.)

Figure 2-11 Pneumatic operator. *(Courtesy Johnson Service Company.)*

2.2.5 Pneumatic Actuators

Sooner or later the output of a pneumatic controller activates a pneumatic actuator which positions a valve or damper. A pneumatic actuator is simply a piston and spring in a cylinder (Figure 2–11). When control air enters the cylinder it drives the piston to compress the spring until the spring pressure and the load on the connecting rod balance the air pressure. The stroke may be limited by adjustable stops. The connecting rod may drive a valve stem directly, or operate a damper by means of a linkage. Different spring ranges are available for sequenced control—for example, full 3 to 13 psi range is most often used but 3 to 8 psi, 8 to 13 psi and other ranges are available.

2.2.5.1 Positive Positioners A standard pneumatic actuator may not respond to small changes in control pressure, due to friction in the actuator or load, or to changing load conditions such as wind acting on a damper blade. Where accurate positioning of a modulating device in response to load is required, positive positioners are used.

A positive positioner (or positive-positioning relay) is designed to provide full main control air pressure to the actuator for any change in position. This is accomplished by the arrangement shown schematically in Figure 2–12. An increase in branch pressure from the con-

Figure 2-12 Positive positioner.

troller (A) moves the relay lever (B), opening the supply valve (C). This allows main air to flow to the relay chamber and the actuator cylinder, moving the piston (not shown). The piston movement is transmitted through a linkage and spring (D) to the other end of the lever (B), and when the force due to movement balances out the control force the supply valve closes, leaving the actuator in the new position. A decrease in control pressure will allow the exhaust valve (E) to open until a new balance is obtained. Thus, full main air pressure is available, if needed, even though the control pressure may have changed only a fraction of a psi. The movement feedback linkage is sometimes mounted internally. Positioners may be connected for direct- or reverse-action.

2.2.6   Relays

Many different types of pneumatic relays are manufactured. Mostly they use some variation of the non-bleed controller shown in Figure 2–2. An external force from another controller or relay replaces the force due to temperature or pressure change. Thus there may be a reversing relay (Figure 2–13) in which a direct-acting control input force may be changed to a reverse-acting output. The output of this and other relays may also be amplified or reduced with respect to input, so that one input to several relays may produce a sequence of varied outputs.

Another type of relay will produce an output proportional to the difference between two inputs (Figure 2–14). Another produces an output equal to the higher (or lower)) of two pressures (Figure 2–15).

Figure 2-13   Reversing relay.

Figure 2-14   Relay: output proportional to difference between two signals.

Two-position relays use the principles illustrated in Figures 2–5 and 2–6.

To illustrate sequencing, envision a controller with a 6 to 9 psi output over the desired control range. It is desired to operate three valves so that one goes from fully open to fully closed with a 3 to 8 psi signal, a second goes from closed to open with a 5 to 10 psi signal and a third has a control range of 8 to 13 psi from closed to open. These are to operate in sequence over the 6 to 9 psi control range. The first valve would utilize a relay which produced a 3 to 8 psi output from a 6 to 7 psi control input.The second relay would provide a 5 to 9 psi output over a 7 to 8 psi input and the third

Figure 2-15   Relay: ouput proportional to higher of two pressures.

would provide an 8 to 13 psi output over the remaining 8 to 9 psi input change.

2.2.7 Master-Submaster Thermostats

The master-submaster principle is used extensively in control systems. It consists of two thermostats, one of which (the master) senses some uncontrolled variable, such as outdoor temperature, and sends its output to the second (the submaster). This has the effect of resetting the set point of the submaster, which is sensing the controlled variable. Adjustments are provided for basic settings of both thermostats and for setting up the reset schedule, that is, the relationship between submaster set point and master output signal. Figure 2–16 shows the system schematically. Both instruments are basically of the non-bleed type. The master may also be a bleed-type controller.

This principle may be extended to utilize a single non-bleed controller with two remote bleed-type sensors as in Figure 2–17. In this case one sensor acts as a master to reset the set point of the other sensor. This arrangement is used extensively in current practice, both for master-submaster control and, with one sensor, for single-point

Figure 2-16   Master-submaster thermostat.

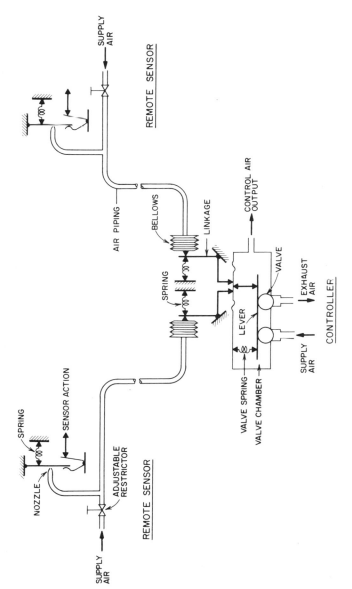

Figure 2-17 Controller with two remote sensors.

sensing and control without reset. It has the advantage that the output of the sensor is independent of the controller action, and can be used to transmit information to other devices, such as supervisory panels or pneumatic thermometers.

2.2.8   Dual-Temperature Thermostats

This classification includes day-night and summer-winter thermostats. Two related temperature settings may be made and the transfer from one to the other is made by changing the supply air pressure. The settings cannot overlap and there is a built-in differential, adjustable or fixed depending on the instrument.

A typical unit with two bimetal sensors is shown in Figure 2–18. The piloting relay is similar to those discussed previously. Two control ports are provided and the bimetals are adjusted to close one port sooner than the other. The port in use is determined by a diaphragm-operated transfer mechanism which is actuated by changing the main air supply pressure; typically from 15 to 20 psi.

This schematic illustrates a different way of providing negative feedback. The control air pressure moves a diaphragm which moves the levers supporting the bimetals.

If both bimetal elements are direct-acting, that is, they bend toward the control port on increase of temperature, then the thermostat functions as a day-night instrument. When used for heating, a lower temperature will be provided at night than during the day. If one of the bimetals is direct-acting and the other is reverse-acting, then the thermostat provides heating-cooling control, for year-round use.

Changeover from one supply pressure to the other may be accomplished in several ways: manually, by means of time clock or programmer, by means of an outdoor thermostat or by means of a sensor which detects supply water temperature in a two-pipe system.

Figure 2-18 Dual-temperature thermostat. (*Courtesy Johnson Service Company.*)

(Obviously, it does little good to call for heating when chilled water is being supplied!)

2.2.9 Transducers

Broadly defined, a transducer is a device for transforming one form of energy to another. Since a peneumatic control system must frequently interface with electrical or electronic devices, transducers are necessary. These may be two-position or modulating.

The PE (Pneumatic-Electric) relay is simply a pressure switch which is adjusted to open or close at some value of the control pressure. Sequenced PE relays are frequently used for capacity control of refrigeration systems, for two-speed fan control and similar functions.

The EP (Electric-Pneumatic) relay is a solenoid air valve, usually three-way to supply or exhaust air to or from the pneumatic control circuit. A three-way valve may also be used to change supply pressure (by changing from one supply source to another.)

Modulating transducers change a modulating air signal to a variable voltage output, in either the electric or electronic range; or a variable electric or electronic signal may produce a varying air pressure output. These will be discussed in the sections on electric and electronic controls.

2.2.10 Air Supply

The air supply for a pneumatic control system must be as carefully designed as the rest of the system. If an adequate source of shop air is already available in the building, it may be used; otherwise a complete compressed air plant must be furnished.

Air consumption can be estimated from use factors for the components as shown by the control manufacturer. Good practice requires that the compressor have a capacity at least twice the estimated consumption. In comparing compressor ratings it should be noted that *displacement* and *capacity* are not the same. Capacity will be from 60% to 80% of displacement.

If the system is very large, it is preferable to have two compressors to improve reliability. Generally air is compressed to 60 to 80 psig, and passed through a drier, a filter and a pressure-reducing valve. It is very important that the air be dry and clean, to minimize maintenance on the components.

2.2.10.1 Air Compressors Compressors usually are reciprocating-type, single-stage, air-cooled. Typically they range in size from ¼ hp to 10 hp, and the smaller units are mounted directly on the receiver tank. Motors are started and stopped by a pressure switch on the receiver, set for 60 to 80 psig with about a 10 psi differential. When two compressors are used, the receivers are crossconnected and a single pressure switch with an alternator controls both motors. The alternator is a sequencing device which provides for running a different motor on each alternate start-stop cycle, thus equalizing motor running time.

2.2.10.2 Dryers and Filters Two methods of drying are used. Refrigerated air dryers cool the air sufficiently to condense out the excess moisture which is then removed automatically by means of a trap, similar to a steam trap.

Chemical dryers are also used, with silica gel as the agent. For small systems an in-line unit may be used, and the chemical must be replaced at regular intervals. For large systems a double unit may be used, with one section being regenerated while the other is in use.

After the air is dried it must be filtered by means of a pipe-line filter. Since the filter will remove both oil and dirt, it must be serviced regularly. A five-micron filter is adequate.

2.2.10.3 Pressure-Reducing Valves The control manufacturer can and does supply the PRV as part of the pneumatic control system. If a dual-pressure system is used then two PRV's will be used, in series, as in Figure 2–19

2.2.10.4 Air Piping and Accessories Air distribution piping systems almost invariably use soft copper tubing with soldered fittings. Compression fittings may be used at equipment connections. Branch piping is commonly ¼ in. outside diameter. Recent improve-

Figure 2-19 Compressed air supply.

ments in nylon-reinforced plastic tubing have led to acceptance of this material, with compression fittings, in accessible places. Specifications should call for a pressure test of the piping system at about 30 psi.

Air gages are a necessity for adjusting, calibrating and maintaining the control elements. Gages should be provided at main and branch connections to each controller and relay and at the branch connection to each controlled device. In this manner the status of the component is readily identifiable.

Linkages of various types are furnished for connecting operators to dampers. Many of these fit only a particular manufacturer's equipment, but many are universally adaptable. Manual valves of various kinds are available. Some of these are for shutoff only. Small needle valves may be used as adjustable restrictors which can be set in infinite steps to provide a desired maximum output pressure at a given flow.

3 Electric and Electronic Control Devices

3.1 ELECTRIC CONTROL DEVICES

Electrical controls are available in a wide variety of configurations, for every conceivable purpose in HVAC applications. All of these are based on one of four operating principles; the switch, the electromagnetic coil and solenoid, the two-position motor and the modulating motor.

It is highly desirable that the reader understand the basic principles of electrical circuits. This book will not discuss these principles, since they are readily available from many sources. The manufacturers' handbooks listed in the bibliography include some very good and simple presentations.

Any electrical circuit includes three elements: a power source, a switch and a load (Figure 3–1). The load represents resistance and

Figure 3-1   Electrical circuit.

Figure 3-2   Bimetal sensor, electric.

consumes the power. The switch serves to turn the power on and off. In an HVAC control system the load will be an actuator or relay, the switch will be the sensor-controller. The power source is usually the building electrical power, which may be used at a normal 120 V or transformed to some lower voltage, typically 24 V. Some electrical devices use direct current (DC). This may be supplied by a battery or from an alternating current (AC) source by means of a transformer and rectifier.

3.1.1   Two-Position Controls

3.1.1.1   Sensors   Any of the sensors described in the chapter on pneumatic controls may be used for two-position (on-off) electric control. Certain modifications may be necessary.

The bimetal is very commonly used in electric thermostats since it can serve to conduct electricity. Figure 3–2 illustrates a simple single-pole, single-throw (SPST) bimetal thermostat. When it is used for heating, a decrease in room temperature will cause the bimetal to bend toward the contact. When the contact is almost closed, a small permanent magnet affects the bimetal enough to cause a quick final closure and lock it in place. This magnet also causes a lag in the release, with resultant quick opening of the contact. This minimizes arcing and burning of the contacts, and eliminates chattering.

The bimetal may also be arranged in a spiral, fixed at one end and fastened to a mercury switch at the other. (Figure 3–3). The mercury switch is simply a glass tube partially filled with mercury and with

Figure 3-3   Mercury switch.

wiring connectors at one or both ends. It is loosely pivoted in the center so that when it turns past center the weight of the mercury running to the low end causes it to pivot farther. The mercury bubble acts as a conductor to connect the electrodes.

The mercury tube may also be used with a bourdon tube pressure sensor.

The diaphragm movement of bulb and capillary sensors can be used to trip an electric switch. Bellows sensors can be used in the same manner. The tripping action can be direct or through a linkage. A bistable spring or over-center mechanism is required to provide the snap action, as described above.

Humidity sensors may also be used to trip switches, through either bending action or expansion.

3.1.1.2 Electromagnetic Devices   Sometimes referred to as electromechanical devices, this class includes relays and solenoid valves, as well as motor starters.

All of these elements utilize the principle of electromagnetism. When an electric current flows through a wire a magnetic field is set up around the wire. If the wire is formed into a coil then the magnetic field may become very strong, and a soft iron plunger placed in proximity to the end of the coil may be drawn up inside it. This is the "solenoid" which can then be used to operate a valve or set of contacts.

Solenoid valves are made in many sizes and arrangements, for control of water, steam, refrigerants and gases. Figure 3–4 shows a typical two-way valve. This valve is held in the normally closed position by fluid pressure. When the coil is energized, the plunger is lifted and opens the valve. Some models are arranged with an internal pilot: a small port which is opened by the solenoid allowing fluid pressure to open the valve. Three-way arangements are common and four or more

Figure 3-4   Solenoid valve.

Figure 3-5   Relay coil and solenoid.

ports are not unusual. The maximum size for a solenoid valve is about 4 in. pipe size. Large sizes lead to problems of pressure and water hammer due to quick opening and closing.

Control relays are designed to carry low-level control voltages and currents, up to about 15 A and 480 V. The contact rating of a control relay will vary with the voltage and with the type of load. The rating will be higher for a resistive load than for an inductive load. A relay can "make" a circuit with a much higher current than it can "break" without arcing and burning the contracts. The "breaking" capacity of the relay should therefore be the criterion.

One typical control relay configuration is the coil and solenoid as shown in Figure 3–5. The figure shows double-pole, double-throw (DPDT) contacts, but numerous arangements from SPST to as many as eight poles are available. The armature is spring-loaded so that it will return to the "normal" position when the power to the coil is turned off. Typically, also, coils are available for most standard voltages, that is, 24, 48, 115, 120, 208, 240, 480.

The solenoid-type relay is also available in "latching" arrangements, so that it may be driven to one position by a short-time energization of the coil and will stay in that position until returned by energizing a second coil. One latching system uses weak permanent magnets at each electromagnetic coil. These permanent magnets are not capable of displacing the solenoid, but will hold it in position (Figure

Figure 3-6   Permanent-magnet latching-type relay.

Figure 3-7   Clapper-type relay.

3–6). Another method uses a mechanical latch or detent, which is tripped when the electromagnetic coil is energized. The advantage of the latching relay is that no power is required to hold it in position. The disadvantage is that it does not "failsafe" when power is removed.

Electromagnetic coils are also used in "clapper-type" relays (Figure 3–7). The coil is mounted near a soft iron bar which is part of a pivoted, spring-loaded contact armature. When the coil is energized the arm is pulled over to close the contact. Several contact circuits may be mounted on a single armature. Double-throw contacts are also supplied. This type of relay usually, but not always, has a lower current and voltage rating than the solenoid type. The units are available as reed relays, and some miniature versions are made for electronics work.

Contactors are similar in all respects to solenoid-type control relays, but are made with much greater current-carrying capacity. They are usually used for electric heaters, or similar devices with high power requirements. A special kind of contactor uses mercury switch contacts, to allow for the frequent cycling required in electric heating applications.

Motor starters also use the solenoid coil actuator, and are similar to relays but with the addition of overload protection devices. These devices sense the heating effect of the current being used by the motor and break the control circuit to the coil if that current exceeds the starter rating. A typical across-the-line starter schematic is shown in Figure 3–8. This subject is discussed further in Chapter 8.

Time-delay relays, as the name implies, provide a delay between the time the coil is energized (on-delay) or deenergized (off-delay), and the time the contacts open and/or close. This delay may range from a small fraction of a second to several hours. Three general classes of time delay relays are available: solid-state, pneumatic and clock-driven.

Figure 3-8   Motor starter.

Solid-state timers use electronic circuits to provide highly accurate, very short delay periods. Timing ranges vary from 0.05 sec to about 60 sec. Pneumatic timers use the familiar solenoid coil principles but the movement of the solenoid is delayed by the diaphragm cover of a pneumatic chamber with a very small, adjustable leakport (Figure 3–9). In the on-delay sequence, when the coil is energized the solenoid pushes against the diaphragm, forcing the air out of the chamber. The leakport governs the rate of escape, and therefore the time delay. When the coil is deenergized, a check valve opens to allow the chamber to refill rapidly. Delays from about 0.1 sec to 60 min are available. Off-delay (after deenergization) is also available.

Clock timers utilize a synchronous clock motor which starts timing when the power is turned on. At the end of the timing period the control circuit contacts are opened or closed. When the power is turned off the device immediately resets to the initial position. Clock timers are available with ranges from a fraction of a second to about 60 h. Many special sequences and capabilities are used in process control.

Most of these time-delay relays may be provided with auxiliary contacts which open and/or close without delay, as in an ordinary relay.

Sequence timers include a synchronous motor which drives a cam

Figure 3-9   Time-delay relay, pneumatic type.

Figure 3-10   Two-position spring-return motor.

shaft, through a chain of reducing gears. Adjustable cams operate switches in any desired sequence. The timer may go through a single cycle, then stop; or it may run continuously, repeating the cycle over and over. Units with as many as 16 switches are available. Modulating or floating type motor control can also be used if needed.

3.1.2   Two-Position Motors

Two-position motors are used for operating dampers or for valves which need to open and close more slowly than a solenoid coil will allow. Motors may be unidirectional, spring-return; or unidirectional, three-wire.

A spring-return motor is shown in Figure 3–10. When the controller closes its SPST switch, the motor winding is energized from A to B. This starts the motor and it runs, driving (through a reduction gear) a crankshaft and linkage to open or close a valve or damper. A cam is mounted on the shaft and at the proper time (usually 180° of rotation but sometimes less) the cam throws the limit switch from B to C. This added coil resistance reduces the current to a "holding" level, which will hold the motor in this position but will not cause damage. When the controller switch opens, the spring returns the motor to its original position.

Figure 3–11 shows a three-wire motor. For discussion purposes assume that this motor is operating a heating valve and the valve is closed in the position shown. A double-throw controller is required, and assume that this is a thermostat which closes contact B on temperature rise and contact A on temperature fall. The controller is shown in the satisfied position. The limit switches (marked SW1, SW2) are operated by a cam on the motor crankshaft. On a fall in room temperature the thermostat closes contact A establishing a circuit through SW2 to the motor field coil. The motor runs and almost im-

Figure 3-11   Two-position motor, three-wire arrangement.

mediately SW1 closes, establishing a maintaining circuit to the coil. Now the motor will run 180° regardless of what happens at the controller. When a 180° stroke has been completed SW2 is opened by the cam and breaks the circuit to the coil, stopping the motor. The valve is fully open. On a rise in temperature the controller breaks contact A and makes contact B, establishing a circuit through SW1 to the coil, and restarting the motor. As the valve starts to close SW2 closes and makes the maintaining circuit. When the valve is fully closed the cam opens SW1, stopping the motor.

In case of power failure the three-wire motor will stop and stay wherever it may be. The spring-return motor will return to normal position.

These motors may be provided with auxiliary contacts, cam-operated, which open and/or close at any desired point in the stroke.

3.1.3   Modulating Motors

Modulating motors are used for proportional and floating controls. They must, therefore, be reversible and capable of stopping and holding at any point in the cycle. The motors used in these devices will be either reversible two-phase induction motors or reversible shaded-pole motors.

3.1.3.1   Reversible Induction Motors   Schematically, the two-phase induction motor (Figure 3–12) has two field windings directly connected at one end (C) and connected at the other by a capacitor. Power may be supplied at A or B on either side of the capacitor, with the other power connection at C. If, for example, alternating current power is connected across A and C, then coil 1 is directly powered and coil 2 is indirectly powered through the capacitor. The effect of the capacitor is to introduce a phase shift between

Figure 3-12   Reversible induction motor.

coils 1 and 2, thus a rotary motion is imparted to the motor armature. If the power is applied across B and C, the phasing is reversed and the motor runs in a reverse direction. This unit may also be called a capacitor or condenser motor, although it is not the same as the capacitor-start, induction-run single-phase motor commonly found on large household appliances.

3.1.3.2 Shaded-Pole Motors The shaded-pole motor is constructed with a main field coil which is directly energized. However, by itself this will not start the motor. To provide a 'biasing" effect shading coils are added as shown in Figure 3–13. These are

Figure 3-13   (Courtesy Barber-Colman Company.)

Figure 3-14   *(Courtesy Barber-Colman Company.)*

powered by transformation effect from the main field coil, and when both ends of the shading coil windings are shorted out, a phase lag is caused in part of the field. This causes a rotating field which starts the motor and improves its efficiency while running. This arrangement is unidirectional. To provide a reversing motor two additional shading coils are added (Figure 3–14) and wired as shown in Figure 3–15. Grounding one pair of shading coils causes the motor to run in one direction, grounding the other pair causes the motor to reverse rotation.

3.1.3.3 Modulating Motor Control Two general types of control configurations are used with modulating motors.

Floating control includes a three-wire controller with a center "dead spot" and no feedback. The motor is geared down to provide a slow change in the controlled device, so that sometimes system response can cause the sensor to return the switch to the center-off position before the motor has traveled the full stroke. Figure 3–16 shows a shaded-pole motor with this type of control. The limit switches are cam-operated at the end of the stroke (as described under two-position motors). The sensor controller is commonly called a "floating controller." This same controller may be connected in a similar manner to an induction motor.

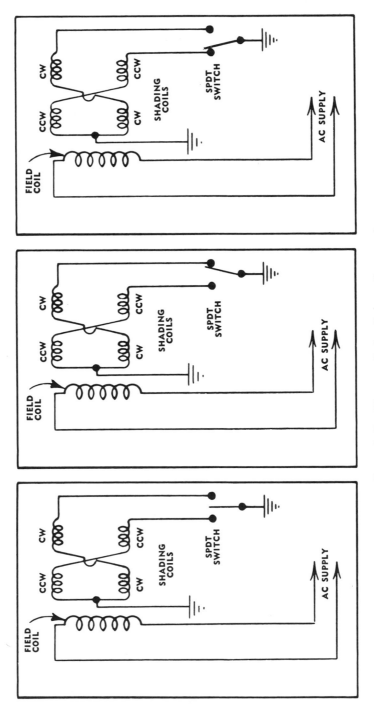

Figure 3-15 (Courtesy Barber-Colman Company.)

Figure 3-16 (Courtesy Barber-Colman Company.)

Figure 3-17   Modulating induction motor.

Fully modulating control requires negative feedback at the motor as in Figure 3–17. Figure 3–18, which shows only the potentiometers and relay coils, will aid in understanding the following description. In the balanced condition shown the wiper arm of the controller potentiometer is centered on its resistance winding, as is that of the feedback potentiometer. Under these conditions the currents through the two coils of the balancing relay are equal, and the relay arm is centered between contacts (Figure 3–18(A)). If the controller responds to temperature (or pressure) change and moves its potentiometer wiper arm, the circuits are unbalanced and the current differences in the two relay

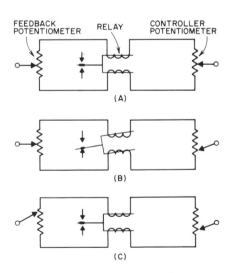

Figure 3-18   (A) Balanced position.   (B) Controller moved, relay unbalanced.   (C) Feedback moved, relay rebalanced.

coils cause the relay arm to swing to one contact (Figure 3–18(B)). This starts the motor and causes it to actuate a valve or damper to offset the variation from set point. The motor operation also moves the feedback potentiometer wiper, which offsets the effect of the controller. When balance is restored (with the motor in a new position) the relay coil currents are again equal, the relay arm is centered and the motor stops (Figure 3–18(C)). The limit switches are needed to stop the motor at the end of the stroke if the sensed conditions are beyond the throttling range of the controller.

The modulating motor may be equipped with cam-operated auxiliary switches which open or close at any point in the stroke. It may also drive an auxiliary potentiometer which provides input to make another modulating motor "follow" the action of the first motor driving another valve or damper.

3.2 ELECTRONIC CONTROL DEVICES

Electronic controls are distinguished from electric controls by the use of low voltages (about 15 V maximum, but 5 V is common), and general use of solid-state or vacuum tube components for amplification and logic. Again, the basic principles involved will not be discussed here, except for bridge circuits (see below). Electronic devices are used primarily as sensors and controller-amplifiers, with relays or transducers to utilize pneumatic or electric actuators.

3.2.1 Bridge Circuits

The original and most commonly used bridge circuit is the "Wheatstone bridge" (Figure 3–19). The bridge is formed by four resistances connected as shown. Power is connected to two "corners"

Figure 3-19 Wheatstone bridge.

Figure 3-20   Wheatstone bridge with remote sensor.

of the bridge and output to the two opposite corners. One or more of the resistances may be variable. (R4 is shown here as variable by the arrow across it.) When all resistances are equal the output is zero.

Then if one or more of the resistances is changed the bridge becomes "unbalanced" and an output signal results which is approximately proportional to the resistance change.

In the ordinary electronic sensor, the variable resistor is the sensing element and is often mounted remote from the rest of the bridge and the amplifier (Figure 3–20). When remote mounting is required, some method must be used to compensate for the length of the conductors, and each control manufacturer has his own method of compensation.

This very simple arrangement does not allow for adjustment of set point, or calibration. For these functions it is necessary to add two more resistances (Figure 3–21). The set-point adjustment is in series (or parallel) with the sensor resistor and the calibration adjustment is a potentiometer with an adjustable wiper arm.

Figure 3-21   Bridge with calibration and set point.

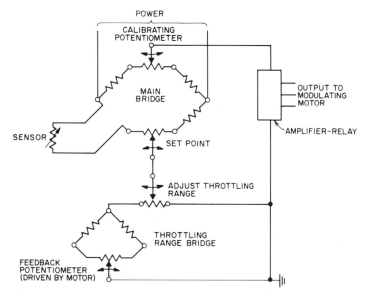

Figure 3-22   Throttling range bridge.

To provide negative feedback for modulating controls a "throttling range bridge" must be added to the circuit. This is wired in series with the main bridge (Figure 3–22). It includes a variable potentiometer with a wiper arm which is driven by the controlled device motor. When the main bridge is unbalanced and causes the motor to run, this potentiometer adjusts the throttling range bridge to offset the effect of the sensor and rebalance the system. In a simpler arrangement, the motor-driven potentiometer may be placed in series with the sensor, but then throttling range is not adjustable.

Sensing and relating two control points, such as room temperature and discharge temperature, may be done with a single bridge circuit (Figure 3–23). Here the two sensors are shown on opposite sides of the bridge. The room sensor has the same basic resistance as the other legs of the bridge. The discharge sensor however usually has only 10% to 20% of the basic resistance, with the balance being furnished by fixed resistors. This gives the discharge sensor less "authority," the amount being expressed as the ratio of discharge sensor resistance to room sensor resistance. Typical values are 1000 ohms for room sensor and 100 ohms for discharge sensor. Then the authority of the discharge sensor is 100:1000 or 10% which means that a 10° rise in discharge temperature is necessary to rebalance the bridge after a 1° fall in room temperature.

Figure 3-23   Bridge with two sensors.

Thus, when room temperature decreases, the resistance of the room sensor R1 is decreased. This unbalances the bridge and causes an output to the amplifier. This may cause a hot water valve to open, which increases the discharge air temperature. This is sensed by the discharge sensor R2 and results in an increased resistance at the sensor until the bridge is rebalanced.

Figure 3–24 will serve to clarify the situation. Figure 3–24(A) shows the bridge in the balanced condition, with a 12 V drop being shared

Figure 3-24   (A) Bridge balanced. (B) Room temperature decrease, bridge unbalanced. (C) Discharge temperature increase, bridge rebalanced.

equally by each of the two resistors in each side. Then there is no difference in potential across A–B; therefore no output. When the resistance of R1 decreases the voltage drop across it will also decrease, say to 5 V. But to maintain the total 12 V drop, the drop across R3 must increase to 7 V (Figure 3–24(B)). Now there is a difference in potential from A to B, which the amplifier will sense and utilize to operate a valve motor. As the discharge temperature increases, resistance R2 will increase until the voltage drop across R2 + R5 equals 7 V (leaving 5 V across R4) and the bridge is rebalanced (Figure 3–24(C)). Output ceases and the valve motor is stopped at this position.

The system as shown is set up for heating. For cooling it is necessary to interchange the location of R4 with that of R2–R5.

3.2.2   Electronic Sensors

Electronic sensors may be bellows or bulb and capillary type, with output of the bellows or diaphragm linked to a potentiometer wiper arm. More often, however, wound wire resistors are used. The wire is temperature-sensitive and resistance increases as temperature increases. These elements are made in various configurations, with and without local reset and calibration. Most often, reset and calibration are performed at the amplifier.

One form of humidity sensor utilizes a hygroscopic chemical applied as a coating between sheets of gold foil. As the chemical absorbs or gives up moisture in response to changes in relative humidity of the surrounding air, the resistance of the system changes. Another form of humidity sensor uses granulated carbon packed between two terminals. As the moisture is absorbed the individual grains of carbon are pushed apart, increasing resistance.

A compressible carbon element can be used as a pressure sensor. An increase in pressure squeezes the carbon grains closer together and decreases electrical resistance.

Thermocouples are seldom used in electronic control but rather for temperature indication purposes. However there is one thermocouple application which is common in small residential gas-fired heating systems. A special form of thermocouple called a thermopile is inserted in the gas burner pilot flame. The heat of the flame generates enough electric current to power a special gas valve, through a bimetal thermostat. Since the small available current will not operate a conventional solenoid valve, the valve used is a balanced diaphragm type. A small solenoid pilot valve opens to allow gas pressure from the supply line to move the diaphragm and open the valve.

3.2.3   Amplifiers and Transducers

As noted above, the output of a bridge circuit is usually sent to an amplifier, since this output may be only a few millivolts. Amplifiers use solid-state or vacuum tube components to amplify this small signal to a level where it is usable by a low-voltage electric actuator. When pneumatic actuators are used with electronic controllers, electronic to pneumatic transducers are required.

An electronic to pneumatic transducer commonly consists of a piloted bleed-type pneumatic relay-controller, as described in paragraph 2.2.3, together with a small servo-motor which is driven by the electric signal from the amplifier. This servo-motor positions the flapper valve at the relay control port, thus varying the controller output pressure in response to the electric signal.

A pneumatic to electric or electronic transducer uses a potentiometer, with the wiper arm driven by a bellows which senses the output of the pneumatic controller. Then the potentiometer resistance varies with branch line pressure.

4 Fluidic Control Devices

4.1 INTRODUCTION

The first use of fluidic principles in control applications occurred about 1960, so it is only in the past two or three years that fluidic control devices have become readily available. Their rapid acceptance, simplicity, reliability and ease of application should contribute to a large increase in their use in the next few years. Although they use lower pressures than pneumatic devices, more air is consumed, so fluidic systems do require more power to operate.

Fluidic devices utilize the dynamic properties of fluids, in contrast to pneumatic and hydraulic devices which depend primarily on static properties. Almost any gas or liquid can be used, though compressed air is the preferred medium. Logically, fluidic devices resemble electronic devices, and the same logic terms are frequently applied.

The principles most commonly used in fluidic devices are wall attachment, turbulence amplification and vortex amplification.

4.2 WALL ATTACHMENT DEVICES

A wall attachment device functions because of the *Coanda effect*: the property of a jet to attach itself to the surface of an adja-

49

Figure 4-1   Elementary air jet.

Figure 4-2   Wall attachment principle.

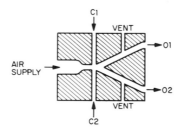

Figure 4-3   Bistable amplifier.

cent plate (or wall). A jet of air flowing from a nozzle entrains air with it (Figure 4–1). If the jet is confined between two parallel surfaces (assumed to be in the plane of the page), then the entrainment takes place only in that plane. If a plate or wall surface is added at one side of the jet (Figure 4–2) air entrainment on that side is reduced. This reduces the air pressure near the wall and the jet is "bent" until it is attached to the wall. If a similar and symmetrical wall is provided on the other side of the jet, then it may become attached to either side with only a slight nudge. Although in theory the shape of the jet is not important, in practice a rectangular jet is found to work best.

This effect can be used to create a logic relay-amplifier with a memory (Figure 4–3). If air is supplied at control port C1 the jet is deflected away from the port and attaches itself to the opposite wall, providing an output signal at O2. If the control signal is removed, the jet will still continue to supply O2. If now a signal is applied at C2 the jet will switch to the opposite wall and the output signal will appear at O1. Amplifications on the order of 3 or 4 to 1 (output to control signal) are possible. The vents are provided to allow continuing airflow if the output is blocked, since interruption of airflow would cause the

Figure 4-4   OR-NOR element (monostable amplifier).

device to operate improperly. This device is called a "bistable amplifier with memory."

A similar but different device is the "OR-NOR" amplifier of Figure 4–4. With no signal at C1 or C2 the output is at O2 (NOR). With a signal at C1 or C2 or both, the output is at O1 (OR). When the signal or signals are removed the output returns to O2.

Figure 4–5 is similar to the bistable amplifier of Figure 4–3 but with the walls cut away just beyond the control ports. With no walls and no control signal, the supply jet goes straight out and produces two equal outputs at O1 and O2. If a control signal is applied at C1 or C2 the jet is deflected in proportion to the strength of the signal, and the outputs at O1 and O2 become unequal. This is a "proportional beam deflection amplifier."

All of the above devices operate best at low pressures: 1 to 5 psig supply and 0.1 psig or less control. Output pressures are usually less than 1 psig.

4.3   TURBULENCE AMPLIFIERS

A jet issuing freely from a nozzle can be adjusted to a fairly long laminar flow pattern (in the order of 1 in. or more). A receiving nozzle in the path of this pattern will then receive some flow and

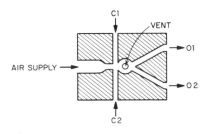

Figure 4-5   Proportional jet amplifier.

Figure 4-6   Turbulence amplifier.

pressure (Figure 4–6). A small control jet directed at the laminar jet will cause it to break up, thus decreasing the pressure felt at the receiver. This change in output is proportional to the control pressure change, with considerable amplification (as much as 10:1). Also, more than one control jet can be used. The turbulence amplifier is a very low pressure device, typically utilizing supply pressures of 10 in. water and control pressures of 0.4 in. water.

4.4   VORTEX AMPLIFIERS

A vortex amplifier has a cylindrical body with supply air introduced at the side and output at one end (Figure 4–7). With no control flow, the air flows directly to the outlet, and flow is not restricted. When a control signal is applied at a tangent to the cylinder wall and at right angles to the supply flow, a swirling action is created, forming a vortex with high resistance and, therefore, lower flow. The control pressure must be higher than the supply flow, and supply output can be reduced, in a proportional manner, to as little as 10% of supply input. The device is most often used as a valve or variable restrictor.

4.5   RADIAL JET AMPLIFIER

A variation of the turbulence amplifier is the radial jet amplifier (Figure 4–8). Two jets are arranged to oppose one another. At the point where they meet a radial pressure area is created, and when

Figure 4-7   Vortex amplifier.

Figure 4-8 Radial jet amplifier.

this is confined in a chamber an output signal can be produced. The diagram shows a reference jet (B) which can be adjusted to a desired pressure (set point), and a signal input jet (A) from the sensor. When A is less than B the radial jet forms in the left-hand chamber and is vented to atmosphere. When A is greater than B then the pressure is developed in the right-hand chamber and an output signal is generated which can power an actuator or reset a relay.

4.6 FLUIDIC TRANSDUCERS

You will have noticed that fluidic control pressures are very small. To operate controlled devices such as valves or motors, it is necessary to provide transducers. Fluidic to pneumatic transducers are most common, since both use air, though at greatly different pressures. All fluidic transducers operate on one of two basic principles: direct force or assisted direct force.

In Figure 4–9 the fluidic signal operates against a diaphragm to deflect it. This *direct force* action could close an electrical switch or move a pilot valve in a pneumatic relay. Even with a large diaphragm the available force is small.

In Figure 4–10 a high-pressure air supply is provided to *assist* or amplify the fluidic signal. With a low fluidic signal the high-pressure air is vented. When the fluidic signal is increased the diaphragm 1 moves to restrict or close the vent nozzle, increasing the pressure on diaphragm

Figure 4-9 Direct force transducer.

Figure 4-10   Assisted force transducer.

Figure 4-11   Three-way valve (with assisted force operation).

2. This higher pressure is adequate for positioning valves or dampers directly. The action can be proportional or two-position, depending on the construction of the diaphragm.

Figure 4–11 shows a three-way two-position pneumatic valve, driven both ways by two assisted fluidic signals operating on a piston. This could also be constructed to use a single signal with spring return. Such a valve has a very fast response time.

Fluidic-electronic transducers are also available. These generally use strain gages or pressure-sensitive transistors to modify the low-power electronic signal. They are not generally used in HVAC systems at this time.

4.7   MANUAL SWITCHES

A simple fluidic pushbutton switch operates on the bleed principle (Figure 4–12). The air supply is vented through an orifice and chamber under the button. When the button is depressed the vent is closed and the output signal increases.

Figure 4-12   Fluidic pushbutton.

Figure 4-13   Selector switch.

Figure 4–13 shows a selector switch which provides a choice of two outputs. This can be built in a rotary or slide arrangement, with a choice of many outlets or combinations thereof.

Any of the manual or automatic air valves commonly used for pneumatic systems can also be used with fluidic controls.

5 Flow Control Devices

5.1 DAMPERS

Dampers are used for the control of airflow to maintain temperatures and/or pressures. Some special types of dampers are used in HVAC equipment, such as mixing boxes and induction units, but the control dampers in air ducts and plenums are almost invariably the "multi-leaf" type. Two arrangements are available: parallel-blade and opposed-blade. As Figure 5–1 shows, in parallel-blade operation all the blades move in the same (or parallel) way. In opposed-blade

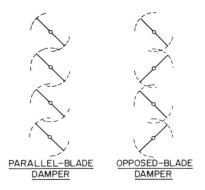

PARALLEL–BLADE
DAMPER

OPPOSED–BLADE
DAMPER

Figure 5-1 Airflow dampers.

operation adjacent blades move in opposite directions. Without going into the physics of the situation it would appear that the opposed-blade damper is superior to the parallel-blade damper for modulating control. This is, in fact, true. Parallel-blade dampers, like flat-seated valves, should be used only for two-position control.

5.1.1 Pressure Drop

As with control valves, there is an optimum relation between control linearity and the initial pressure drop in the system. For a typical parallel-blade damper to achieve essentially linear modulating control it is necessary that the pressure drop through the damper in the full open position be about 50% of the total system pressure. This is shown in Figure 5–2 (curve K) and Table 5–1 as a typical flow-vs-stroke curve for these conditions. To accomplish an equivalent linearity with a typical opposed-blade damper, the initial pressure drop through the damper need be only about 10% of the system pressure drop. See Figure 5–3 (curve F) and Table 5–1.

It is then obvious that the opposed-blade damper is preferable from a noise and energy-loss standpoint.

Figure 5-2 Parallel-blade damper flow characteristics. (Redrawn from E. J. Brown, *Heating, Piping and Air Conditioning,* April 1960, p 171.)

TABLE 5–1   Ratios of Damper Resistance to System Resistance Which Apply to Flow Characteristic Curves of Figures 5-2 and 5-3[a]

| Parallel-leaf Dampers | | Opposed-leaf Dampers | |
Open Damper Resistance. Percent of System Resistance	Flow Characteristic Curve	Open Damper Resistance. Percent of System Resistance	Flow Characteristic Curve
0.5– 1.0	A	0.3– 0.5	A
1.0– 1.5	B	0.5– 0.8	B
1.5– 2.5	C	0.8– 1.5	C
2.5– 3.5	D	1.5– 2.5	D
3.5– 5.5	E	2.5– 5.5	E
5.5– 9.0	F	5.5–13.5	F
9.0–15.0	G	13.5–25.5	G
15.0–20.0	H	25.5–37.5	H
20.0–30.0	J		
30.0–50.0	K		

[a] From E. J. Brown, *Heating, Piping and Air Conditioning*, April 1960, p 171.

Figure 5-3   Opposed-blade damper flow characteristics. (Redrawn from E. J. Brown, *Heating, Piping and Air Conditioning*, April 1960, p 171.)

5.1.2   Leakage

Unlike single-seated control valves, dampers are inherently leaky. The amount of leakage at close-off is a function of the damper design. Minimizing leakage increases cost, so we should always look at system requirements and select the damper accordingly. Most manufacturers publish leakage ratings. A simple inexpensive damper may have a leakage of 50 cfm per square foot of face area at 1.5 in. of water pressure, while a very carefully designed and constructed damper may leak as little as 10 cfm per square foot at 4 in. of water pressure.

5.1.3   Operators

Damper motors may be pneumatic, electric or hydraulic. They must have adequate power to overcome bearing friction and air resistance, and in the case of tight-fitting low-leakage dampers, to overcome the "binding" friction of the fully closed damper.

5.1.4   Face and Bypass Dampers

Face and bypass dampers are frequently used with preheat coils, and less often with direct-expansion cooling coils. Figure 5–4 shows a typical preheat installation. The preheat coil is sized with a face velocity based on the manufacturer's recommendations; and with all the air flowing through the coil. The open face damper will then have this face velocity. The face damper and coil pressure drops can then be determined from manufacturer's information (see also reference 5 in the Bibliography). Assume a bypass damper size and determine its pressure drop when full open and with all the air flowing through it. This drop should be approximately equal to the sum of the face damper and coil pressure drops. Adjust the bypass size until

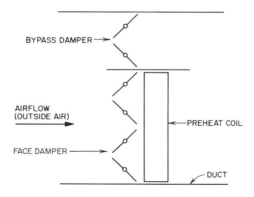

Figure 5-4   Face and bypass dampers.

equality is achieved. For the sake of economy, standard damper sizes should be used if possible.

5.2   STEAM AND WATER FLOW CONTROL VALVES

The proper selection and sizing of valves for the control of steam and water flow requires an understanding of both the valve characteristics and the system in which it is to be used. This section will therefore discuss valve types and sizing methods.

The size of a heat exchanger or coil and the fluid flow rate through it must be based on some maximum design load. But, in practice, the equipment usually operates at part load, so the valve must control satisfactorily over the whole range of load conditions.

The rate of flow through a valve is a function of the pressure drop or head across the valve, in accordance with the hydraulic formula

$$V = K \sqrt{2gh}$$

where: V = fluid velocity in the valve

K = a constant which is a function of valve design

g = acceleration due to gravity

h = pressure drop across the valve.

The change in pressure drop and flow in relation to stroke, lift or travel of the valve stem is a function of the valve plug design. Different types of valve plugs are required to accommodate different control methods and fluids.

5.2.1   Two-Position Valves

These should be of the flat-seat or quick-opening type shown in Figure 5–5. The accompanying graph of percent flow vs percent lift (Figure 5–6, curve A) shows that nearly full flow occurs at about 20% lift. Two-position valves should be selected for a pressure drop of 10% to 20% of the piping system pressure differential, leaving the other 80% to 90% for the heat exchanger and its piping connections.

Figure 5-5   Quick-opening valve. (*Courtesy Barber-Colman Company.*)

Figure 5-6 *(Courtesy Barber-Colman Company.)*

5.2.2 Modulating Valves

The stroke or "lift" of a modulating valve actuator will usually vary directly with the change in the load. If the valve is designed with a linear characteristic, as in Figure 5–7 and curve B in Figure 5–6, then flow rate will vary directly with lift or nearly so. By varying the angle of this V-port arrangement, curves above and below the linear relation can be obtained. There is a minimum flow which is obtained immediately upon "cracking" the valve open. This is due to the clearances required to prevent sticking of the valve. This minimum flow is generally on the order of 3% to 5% of maximum flow and is defined as the "turndown ratio" of the valve. At 5% the turndown ratio would be 20:1.

A linear characteristic valve is excellent for proportional control of steam flow since the heat output of a steam heat exchanger is directly proportional to the steam flow rate. This is because the steam is al-

Figure 5-7 Linear or V-port valve. *(Courtesy Barber-Colman Company.)*

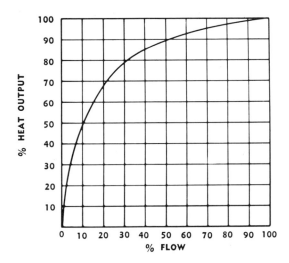

Figure 5-8 Flow vs. heating capacity in a hot water heat exchanger. *(Courtesy Barber-Colman Company.)*

ways at the same temperature, and the latent heat of condensation varies only slightly with pressure change.

Hot water, however, creates a different requirement, since reduction in flow will be accompanied by an increase in the temperature change of the water. The net result may be only a small reduction in heat exchange for a large reduction in flow. Figure 5–8 shows capacity vs flow for a typical hot water coil with 200° entering water and 50° entering air and a design water temperature drop of 20° at full load. This shows that a 50% reduction in water flow will cause only about a 10% reduction in heat output, with an increase to a 40° drop in the water temperature. In order to cut the capacity by 50% the water flow must be reduced by about 90%.

In order to get a better relationship of lift vs output for this case, an "equal percentage" valve is used (Figure 5–9 and curve C of Figure 5–6). The equal percentage valve has a plug shaped so that flow varies as the square root of lift. In order to take full advantage of this con-

Figure 5-9 Equal percentage valve. *(Courtesy Barber-Colman Company.)*

figuration a long stroke or "high lift" must be used (greater than ½ in.). The graph shows a plug with a square root curve. This type of valve should be used for all hot water heating control.

Chilled water, with its smaller temperature differences, can be adequately controlled by a linear valve.

5.2.3 The Flow Coefficient

Most manufacturers publish valve capacity tables based on the flow coefficient C_v. This is defined as the "flow rate in gallons of 60° F water that will pass through the valve in one minute at a one-pound pressure drop." The flow rate at pressure drops other than one pound is found by the formula:

$$GPM = C_v \sqrt{PD}$$

where PD is the desired pressure drop across the valve.

Valve capacity tables usually show C_v and then flow rate at various pressure drops.

5.2.4 Valve Pressure Drop

To determine the required pressure drop for a modulating water valve at full design load it is necessary to look at the entire system. Three conditions are possible:

1. Throttling or closing off any individual control valve has little or no effect on the pressure differential from supply main to return main.
2. There is only one control valve, so that throttling or closing off this valve changes the flow correspondingly in the entire piping system.
3. Other conditions between these two extremes.

In the first case the heat exchanger and related piping between supply and return mains can be considered as a subsystem (Figure 5–10(A)). First consider the subsystem without a control valve. Now if somehow the pressure differential between supply to return mains is varied, the flow rate in the subsystem will vary exponentially as shown in Figure 5–10(B). But a constant differential pressure between supply and return mains has been assumed, with a control valve as shown in Figure 5–11(A). The valve presssure drop at design flow should be at least 50% of the total available head. In order to provide adequate

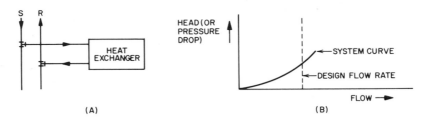

Figure 5-10   A heat exchanger subsystem.

Figure 5-11   Subsystem with control valve.

control at all conditions the major portion of the pressure drop must be taken by the valve. Otherwise the subsystem pressure drop would govern during part of the stroke.

If the valve pressure drop is 50% of the total head then Figure 5–11(B) shows the result. This is the same curve as before for the subsystem without the valve, but the supply to return main differential has been doubled to allow for the valve loss at design flow. Now, as the valve throttles to reduce flow the subsystem pressure drop decreases, and at some flow rate B the valve pressure drop must have increased to compensate.

In the second case exactly the opposite condition prevails. With only one control valve our system would look schematically like Figure 5–12. For this system, flow is again related to pressure drop by an exponential curve as in Figure 5–10(B). A typical pump performance curve of head vs flow will look like Figure 5–13. With the control valve full open these two curves will establish some operating flow rate A as in Figure 5–14. As the control valve modulates toward the

Figure 5-12   A simple system with one control valve.

Figure 5-13 Typical pump curve.

Figure 5-14 Simple system at design and reduced flow rates.

closed position, flow decreases to *B*. But the pump head must increase as flow decreases. This means simply that the control valve pressure drop has had to increase more for the same flow reduction than in the first case we considered. That is, the valve will need a longer stroke or a different plug characteristic at the same stroke in order to obtain the same flow reduction as in the first case.

Cases which fall between these two extremes result from small systems with only a few control valves or from systems in which a few control valves handle most of the water flow. Then any one of these control valves will have a noticeable but not completely governing effect on the pump pressure and flow.

These last two cases can be avoided by the use of three-way valves or pressure-controlled bypass valves, which will maintain the total flow rate (and therefore the supply to return main differential pressure) essentially constant.

5.2.5 Shutoff Head and Static Head

Any flow control valve will have a flow direction symbol on the outside of the body. For a modulating valve the direction of flow should always be such that the flow and pressure tend to hold the valve open (Figure 5–15). As the valve modulates to near the closed position the flow velocity around the plug becomes very high. If the flow and pressure were arranged to help close the valve, then, at some point near closing, the velocity pressure would overcome the spring

FLOW →

Figure 5-15   V-port valve.

resistance and force the valve closed. Flow would then stop, with velocity pressure going to zero. But the static pressure differential alone would not be enough to hold the valve closed, so it would open and the cycle would be repeated. This causes "chatter" and is usually a result of installing the valve backward.

When the valve is properly installed, the closing force must overcome both velocity and static pressure. When the sum of these two pressures reaches a maximum, it is known as the close-off pressure and is one criterion for valve selection.

The static pressure rating of a valve is the highest pressure the valve body will stand wtihout damage or leaking.

5.2.6   Three-Way Valves

Three-way valves are used to provide constant flow through a heat exchanger, whether it be a coil, chiller, condenser or heater. Later chapters will show typical applications.

Figure 5-16   Three-way mixing valve. (Courtesy Barber-Colman Company.)

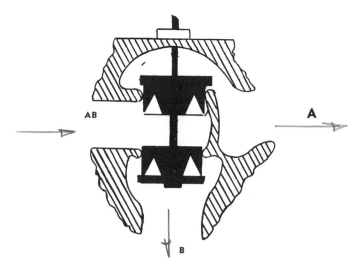

Figure 5-17 Three-way diverting valve. *(Courtesy Barber-Colman Company.)*

Three-way valves are made in two types: mixing and diverting. The internal designs are different (see Figures 5–16 and 5–17), the difference being necessary so that the valve will seat against flow, as discussed in the section on shutoff head above. The "mixing" valve uses a typical linear V-port plug with an added taper on top to seat in the second inlet port. The "diverting" valve uses two V-port plugs which seat in opposite directions and against the common inlet flow. It can be seen that using either design for the wrong service would tend to cause chatter.

Although the valve connections will be marked "A," "B" and "AB," for logic purposes they are usually designated "normally closed," "normally open" and "common." Either the "A" or the "B" connection may become normally open, depending on the arrangement of the motor operator.

5.2.7 Valve Operators

Valve operators may be pneumatic, electric or electronic motors. Such motors have been described in Chapters 2 and 3. Valves may also be "self-contained"; that is, the operator-motor will derive its power from the expansive force of a thermally sensitive fluid in a bulb and capillary tube arrangement (Figure 5–18). Another operator, which is technically electronic, is actually hydraulic, and contains an

Figure 5-18   Self-contained valve. *(Courtesy Penn Controls, Inc.)*

oil reservoir and a tiny pump which pumps oil from one side of a diaphragm to the other.

5.2.8   Criteria for Valve Selection

Flow control valves should be carefully selected to match the characteristics of the system which they are to control. Necessary criteria include flow rate, pressure drop, close-off pressure, static pressure, and type of action. The type of operator used should be compatible with the rest of the control system.

5.3 SUMMARY

The preceding discussion of control devices, while brief, is comprehensive enough to provide an understanding of the variety of elements available and their principles of operation.

The following chapters show how these elements are combined to make systems and subsystems for control of HVAC functions. In most cases any type of energy may be used, the criteria generally being convenience, economics and reliability.

6 Elementary Control Systems

6.1 INTRODUCTION

The preceding chapters discussed the various control elements or units, and how they function with different types of energy. The next four chapters will consider the application of these units to form combinations or "control systems" for the control of HVAC. (The word "system" becomes somewhat overworked here, applying as it does to both "control systems" and "HVAC systems." The difference is always clear in context.) This chapter will consider only small segments of the larger overall systems. Each of these is a complete control system within itself, and all large systems are made up of complexes of these elementary systems.

The following discussions will deal with function only, and except in a few cases, assume that the function can be accomplished by means of any type of energy—electric, pneumatic, electronic or fluidic.

6.2 OUTSIDE AIR CONTROLS

Before deciding on how to control the amount of outside air it is necessary to determine how much is required by the HVAC system and why. For example, certain areas such as laboratories and special manufacturing processes may require 100% exhaust and make-up. Commercial buildings require only minimum ventilation rates. Clean rooms require that a positive internal pressure be maintained to prevent infiltration from surrounding areas, while spaces such as chemical labs and plating shops require a negative pressure to prevent exfiltration.

Once the criteria have been determined, one of the following methods of control can be selected.

6.2.1 Minimum Outside Air

By far the simplest method of outdoor air control is to open a "minimum outside air" damper whenever the supply fan is running (Figure 6–1). This provides the air required for ventilation or exhaust makeup with no complications and no frills. It was one of the first ideas developed and is still used extensively.

6.2.2 Economy Cycle Outside Air

It became apparent, however, that when nominal and fixed amounts (10% to 40%) of outside air are used, there are many times when it is necessary to operate the cooling coil even when outdoor air temperatures are near or below the freezing mark. This gave rise to the so-called "economy cycle" (Figure 6–2), with outside return and relief dampers controlled by air temperature. With outside air at design winter temperature the outside air damper and relief dampers are in minimum open positions (as determined by ventilation and exhaust requirements) and the return air damper is correspondingly in maximum open position. As outside air temperature increases, the mixed

Figure 6-1 Outside air; two-position control.

Figure 6-2   Outside air; economy cycle, no minimum.

air thermostat (T1) gradually opens the outside air damper to maintain a constant low-limit mixed air temperature. Return and relief dampers modulate correspondingly. At some outside temperature, usually between 50° and 60° F, 100% outside air will be provided and used for cooling. As the outside air temperature continues to increase, at 70° to 75° F an outdoor air high-limit thermostat (T2) is used to cut the system back to minimum outside air, thus decreasing the cooling load. This system is extensively used today. Notice the interlock from the supply fan. It is provided in all outside air control systems so that the outside air damper will close when the fan is off.

Figures 6–3 and 6–4 show two alternative methods of economy cycle control. Figure 6–3 is the older arrangement, with a fixed minimum outside air damper which opens whenever the fan runs and is not affected by the temperature controls. The rest of the system operates exactly as described previously.

Figure 6-3   Outside air; economy cycle, fixed minimum.

Figure 6-4   Outside air; alternate economy cycle.

Figure 6–4 utilizes the high-limit controller in a different way. As before, with outside air at minimum design temperature, the low-limit mixed air thermostat (T1) is in control and outside air is at a minimum. As outdoor temperature increases, the outside air damper gradually opens until, when outside air temperature reaches the low-limit temperature set point, 100% outside air is provided. This condition is maintained until the outside air reaches the high-limit temperature setting. Then the high-limit controller takes over, the outside air damper starts closing and continues to modulate toward the closed position as outdoor temperature increases. This control system differs from the other two in two ways: It is not as economical with higher outdoor temperatures; and if the building itself at startup is warmer than the high-limit set point and the outside air, then the high-limit control will have the opposite effect to that intended. Admittedly, this last is a rare case, but it has occurred!

6.2.3   Enthalpy Control

It can be shown that outside air "economy cycle" control based on dry-bulb temperatures (as described in paragraph 6.2.2) is not always the most economical. That is, in very humid climates the total heat (or enthalpy) of the outside air may be greater than that of the return air even though the dry-bulb temperature is lower. For example, in Figure 6–5, the psychrometric chart shows outside air nearly saturated at 70° D.B., while return air at 80° D.B. but much drier, has a lower enthalpy. Since the cooling coil must remove the total heat from the air to maintain the desired condition, it is more economical in this case to hold outside air to a minimum. To measure enthalpy it is necessary to sense two characteristics of the air–dry bulb plus either wet

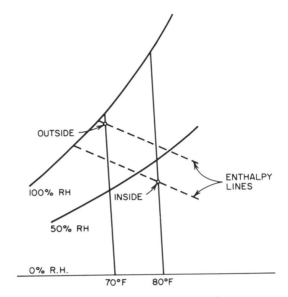

Figure 6-5   Partial psychrometric chart.

bulb, relative humidity, or dewpoint. Several manufacturers now have instruments which will sense dry bulb plus dew point, determine enthalpy, and do this simultaneously for return and outside air, providing an output to control the dampers (see Figure 6–6).

Figure 6-6   Differential enthalpy control.

Figure 6-7   Outside air; static pressure control.

6.2.4   Static Pressure Control

For those spaces requiring a constant positive or negative pressure with respect to their surroundings, the outside, return and relief air dampers will be controlled by static pressure controllers. This will usually be part of a larger system as described later in this chapter. In its simplest form (Figure 6–7), the static pressure controller senses the difference in pressure between the controlled space and a reference location (either adjacent to the controlled space or outdoors) and adjusts the dampers to maintain that pressure differential. The amount of outside air provided must be sufficient to make up any exhaust and to pressurize the space. Proportional control is required. A delaying device or "snubber" is desirable to prevent surges due to sudden changes, such as caused by opening a door.

Chapter 9 considers a more complex push-pull system where static pressure control is required on both supply and exhaust.

6.3   HEATING

Heating in HVAC systems is usually provided by steam or hot water coils with remote boilers. Electric heating coils and direct gas-fired duct heaters are also used, and are discussed in other sections of this book.

Heating may be done as preheat of outside air or mixed air, reheat for humidity control, individual zone control or what we might call "normal" heating. Each of these has its own special control requirements.

6.3.1 Preheat

Preheating is used with large percentages of outside air to prevent freezing of downstream heating and cooling coils and to provide a usable mixed air temperature for dual-duct or multizone systems (if the cold-duct temperature is too low the room control may be unstable). The principal problem in preheating is freezeup of the preheat coil itself. Several methods are used to prevent this.

Figure 6–8 shows the simplest. This is a two position valve in the steam or hot water supply with an outdoor thermostat which opens the valve whenever the outdoor temperature is below 35° or 40° F. (This, incidentally, is an open-loop control.) The filter is downstream of the coil to prevent snow loading in severe winter storm weather. Since no control of leaving air temperature is provided, the preheat coil must be carefully selected to prevent overheating at, say, 30° outside, while still providing adequate capacity at perhaps –10° or –20° F outside design conditions. This is obviously difficult, if not impossible, so a compromise is worked out.

Face and bypass dampers are added at the coil and controlled by means of a downstream thermostat (T2, Figure 6–9) to provide a usable mixture temperature. The difficulty here is stratification of the two air streams. There have been cases where a downstream cooling coil has been frozen by a bypass air stream while the preheat coil was in full operation. The preheat coil should always be located in the bottom of the duct, and, even so, it is desirable to provide mixing baffles. Given sufficient distance in which to provide adequate mixing, this system works well.

Figure 6-8   Preheat; outside air thermostat.

Figure 6-9   Preheat; face and bypass dampers.

In many cases such distance is not available. The best solution in this case is to use hot water with a recirculating pump (Figure 6–10). Now there can be full flow through the coil at all times with the temperature of the water varied to suit requirements. No air is bypassed so there are no mixing problems. Very accurate control of air temperature is possible. Notice the "opposed" flow arrangement with the hot water supply entering the "air leaving" side of the coil.

When dealing with freezing air certain precautions are necessary. For hot water, it has been shown experimentally that water velocities of 2½ to 3 ft/sec in the coil tubes are sufficient to prevent freezing at outdoor temperatures down to −30° F or somewhat lower provided some hot water is being added. If it is necessary to contend with −40° F or below the use of direct-fired systems, gas, oil or electric is recommended. Glycol solutions may also be used.

Figure 6-10   Preheat; secondary pump and three-way valve.

Figure 6-11   Preheat coil with circulating pump.

Steam coils in freezing air should be of the double-tube distributing type with a good slope or vertical arrangement to drain condensate, adequate trap capacity and vacuum breakers. Even then, problems may occur if steam flow is modulated. Traps and drain lines must be insulated if exposed to freezing air.

An alternative pumping arrangement is shown in Figure 6–11. This allows the use of a straight-thru valve. The pump head and horsepower may be somewhat less than in the three-way valve arrangement. (See bibliography reference 22 for a full discussion of comparative performances of these two arrangements.) It should be noted here and remembered in the discussions which follow that in some situations the straight-thru valve may be preferred over a three-way valve. This is particularly true in large systems with multiple chillers and for boilers and circulating pumps. Three-way valves would require that full pumping capacity always be available, even at light loads. With straight-thru valves the flow is reduced at part loads and some of the pumps and chillers can be shut off, with a saving in energy consumption. Three-way valves should be used on coils near the end of the piping system

Figure 6-12   Bypass valve.

to prevent stagnation. In addition, large systems sometimes use a bypass valve between supply and return mains in the chiller/boiler room (Figure 6-12). This valve is controlled by a differential pressure controller to maintain some fixed minimum supply-to-return pressure. As the air handler unit valves operate, the bypass modulates. It should be sized for the capacity of one chilled water/hot water pump, and provided with limit switches to turn on pilot lights when near full open or full closed. Then this sub-system serves to tell the building engineer to start or stop another pump to match the load more closely.

6.3.2   Normal Heating

"Normal" heating refers to the coil in a single-zone, multizone or dual-duct air system which handles all or a major portion of the system air at entering temperatures of 45° to 50° F or higher. In the case of a single-zone unit (Figure 6-13) the supply valve is controlled by a room thermostat (T1), frequently with a high-limit discharge thermostat (T2) added. The discharge thermostat acts as a feedback control, and decreases the operating differential. In dual-duct or multizone systems the supply valve is controlled by a hot plenum thermostat (Figure 6-14). To improve overall controllability, it is desirable to add outdoor reset, decreasing hot plenum temperature as outdoor temperature increases.

If a recirculating pump system is used as in Figures 6-10 or 6-11, one coil may sometimes serve for both preheat and normal heating in a single-zone air handler. The choice will depend on the degree of control required, and the economics.

Figure 6-13   Heating, single-zone.

Figure 6-14   Heating, multizone or dual-duct.

6.3.3   Reheat

Reheat is used for humidity control or individual zone control. In either case, control of steam or hot water supply valves is usually by room thermostat, sometimes in series with a supply duct high-limit thermostat.

6.4   COOLING COILS

Cooling coils are generally confined to the air handling unit, although occasionally recooling coils are required, as, for example, with chemical dehumidifiers. There are two types: direct-expansion (DX) coils and those using chilled water.

6.4.1   Direct-Expansion Coils

DX coils must, by their nature, use two-position control with its inherently wide operating differential. Nonetheless, this system is often used, particularly in small units and where close control is not required. Figure 6–15 shows a typical DX coil control. The room

Figure 6-15   Direct-expansion cooling; two-position control.

Figure 6-16   Direct-expansion cooling; face and bypass control.

thermostat opens the solenoid valve, allowing refrigerant liquid to flow through the expansion valve to the coil. The expansion valve modulates in accordance with its setting to try to maintain a minimum refrigerant suction temperature. A low-limit discharge thermostat, T2, keeps the supply air temperature from becoming too cold.

Controllability can be improved by providing face and bypass dampers (Figure 6–16), but this may lead to such complications as lack of humidity control and coil icing at high bypass rates. It follows that maximum bypass rates must be established and the system may not provide adequate control at very light loads.

A different approach adds a variable back-pressure valve in the refrigerant suction line, controlled by the room thermostat (Figure 6–17). As the room temperature decreases the valve is throttled, increasing the suction temperature at the coil and decreasing the coil capacity. A reversing relay allows the back-pressure valve to be normally open, a necessary condition when the solenoid valve is first opened.

Figure 6-17   Direct-expansion cooling; suction pressure control.

Figure 6-18   Cooling; chilled water, three-way valve.

6.4.2   Chilled Water Coils

Chilled water coils are controlled in much the same way as heating, with three-way or straight-through valves, modulating or two-position. Generally, however, it is desired that cooling coil control valves fail to the closed position since this allows the use of direct-acting controllers. The three-way valve arrangement would then appear as in Figure 6–18 or, if a recirculating pump is used, as in Figure 6–19.

The recirculating pump arrangement is very useful in two cases: (1) for extremely accurate temperature control, and (2) to avoid freezing in those situations where system geometry may make it impossible to avoid stratification of mixed or partially preheated air.

6.4.3   Parallel and Counter Flow

You will notice in both the preceding figures that water flow is shown counter to airflow. This was mentioned before in connection with hot water heating coils, and it is very important in terms

Figure 6-19   Cooling; chilled water, three-way valve and circulating pump.

of heat transfer efficiency. Consider the cooling coil with air flowing through it, decreasing in temperature from 80° to 55° and with water flowing parallel to the air and increasing in temperature from 40° to 50°. On a graph of temperature vs distance this process would appear as in Figure 6–20(A). For counter flow Figure 6–20(B) applies. From heat transfer theory we learn that the heat transfer from air to water in a coil is a function of the K factor through the tube wall and air and water films, the total internal tube surface area and the MED (mean equivalent temperature difference). Given a fixed coil size and fixed rates of air and water flow, the only difference between parallel and counter flow arrangements is the MED. The MED can be calculated mathematically by means of equation (6–1).

$$MED = \frac{GTD - LTD}{\ln \dfrac{GTD}{LTD}} \qquad (6\text{–}1)$$

where: GTD = greatest temperature difference between water and air
LTD = least temperature difference between water and air
MED = mean equivalent temperature difference
ln = natural log (log to base e)

If we calculate the MED for each of the two flow arrangements, we get:

1. Parallel flow: $GTD = 80 - 42 = 38$
   $LTD = 55 - 52 = 3$

$$MED = \frac{38 - 3}{\ln \dfrac{38}{3}} = 13.8$$

Figure 6-20   Chilled water flow in a cooling coil. (A) Parallel flow. (B) Counter flow.

2. Counterflow: $GTD = 80 - 52 = 26$
$LTD = 55 - 42 = 13$

$$MED = \frac{26 - 13}{\ln \frac{26}{13}} = 18.7$$

This is an increase of nearly one-third in the heat transfer capacity, which of course cannot occur without a change in temperatures or flow rates. In practice, it can be shown that, in this example, to keep the same flow rate, capacity and air temperatures, the water temperature can be increased by 5° in going from parallel flow to counter flow, without any decrease in capacity. This would allow water temperatures of 47° in and 57° out, and would actually make the design condition (55° leaving air) impossible with parallel flow. Obviously, the higher water temperature increases chiller efficiency and capacity.

The MED equation also holds when used with a condensing or evaporating fluid. But now counter flow does not apply, since condensing steam or evaporating refrigerant provides essentially a constant temperature in the coil.

6.5   HUMIDITY CONTROL

At different times and for different reasons it may be necessary to raise or lower the humidity of the supply air in order to maintain selected humidity conditions in the air conditioned space.

6.5.1   Air Washer

Consider first the air washer (Figure 6–21). Often used for its sensible cooling capability, it is then known as an evaporative

Figure 6-21   Evaporative cooling (air washer).

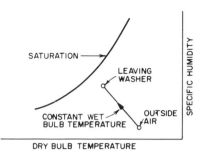

Figure 6-22 Psychrometric chart for evaporative cooling.

cooler. Whether an inexpensive wetted-pad residential-type unit or a large industrial unit with an elaborate system of sprays and eliminators, any air washer operates on the adiabatic cooling principle. That is, the cooling is accomplished by using the sensible heat of the air to evaporate water. Thus, the air passing through the washer changes conditions along a constant wet bulb line, with the final state being dependent on the initial state and the saturation efficiency of the washer (generally 70% to 90%). There is no control of humidity. This is shown in the psychrometric chart of Figure 6–22.

6.5.1.1 Two-Stage Evaporative Cooling Two-stage evaporative cooling may be used as an alternative to mechanical refrigeration, when outdoor conditions allow. This system provides lower dry bulb temperatures and relative humidities than can be obtained with ordinary evaporative cooling. Figure 6–23 shows the arrangement and control of a two-stage evaporative cooling system. Figure 6–24 is the

Figure 6-23 Two-stage evaporative cooling.

Figure 6-24   Psychrometric chart for Figure 6-23.

psychometric chart of the cycle. The dashed lines on the chart show a single-stage cycle as in Figure 6–22.

The cooling tower and precooling coil constitute the first stage. This sensible cooling reduces both the wet and dry bulb temperatures of the air, so that in the second stage a lower dry bulb temperature may be obtained. The room thermostat will be a two-position type, with two-stage control optional. The controllability of the system depends on the condition of the outside air.

6.5.2   Air Washer with Preheat

About the only control that can be applied to the ordinary air washer is to turn the spray water (or pump) on or off. If a minimum humidity is required, it is sometimes necessary to preheat the air to the desired wet bulb temperature. Such a control system is shown in Figure 6–25. The room humidistat senses low humidity and turns on

Figure 6-25   Minimum humidity; air washer with preheat.

Figure 6-26 Psychrometric chart for Figure 6-25.

the spray pump and then opens the preheat coil supply valve. As room humidity increases, the preheat valve is first closed, then if the increase continues the sprays are shut down. Under high outdoor humidity conditions cooling capacity is limited. Final room temperature control is provided by reheat coils. The psychrometric chart in Figure 6–26 shows this cycle.

6.5.3 Air Washer with Preheat and Refrigeration

Obviously, such a system will not control the upper limits of humidity. To accomplish this two choices are available, both requiring refrigeration.

1. Heat and/or cool the spray water by means of shell and tube heat exchangers (Figure 6–27). By allowing the humidistat to control the heat exchanger supply valves it is possible to obtain very accurate control of the humidity as well as the air temperature

Figure 6-27 Humidity control by heating or cooling spray water.

Figure 6-28 Psychrometric chart for Figure 6-27.

leaving the washer. Figure 6–28 is the psychrometric chart diagram of this process, and shows clearly why the reheat coil is still needed.

2. Add a cooling coil in the air washer, creating a "sprayed-coil dehumidifier" (Figure 6–29). The term is not incorrect, since at high humidity conditions dehumidification does take place, but it also acts as a humidifier if necessary. Since the addition of the coil increases the saturation efficiency to between 95% and 98%, it is possible to do away with the humidistat and use a simpler, less expensive dry bulb thermostat (T1) in the air leaving the coil (the so-called "dew-point" thermostat). This is set to maintain a fixed condition of dry bulb and relative humidity. Preheat will be required at low mixed-air humidities. When the mixed-air wet bulb is above the required wet bulb temperature of the air leaving the coil, then refrigeration must be used. Reheat is necessary for final space temperature control. If all control settings are properly selected, it is possible to maintain very accurate control of space temperature and humidity. Figure 6–30 shows the cycle on a

Figure 6-29 Humidity control, sprayed coil and preheat.

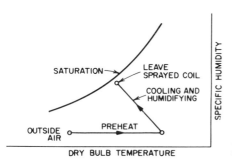

Figure 6-30 Psychrometric chart for Figure 6-29 (winter).

psychrometric chart, assuming cool, low-humidity outside air. The dew-point thermostat senses a decrease in coil leaving-air temperature and shuts down the chilled water flow to the coil and then opens the preheat coil valve. Evaporative cooling is used. As the coil leaving-air temperature increases, the preheat coil valve is gradually closed. A further increase in the controlled temperature above the thermostat setting will cause the chilled water valve to open, using refrigeration for cooling. Figure 6–31 is a psychrometric chart showing what happens with high-temperature outside air. So long as the entering-air wet bulb is above the dew-point thermostat setting, chilled water will be required and the coil leaving-air temperature will be maintained by varying the flow of chilled water through the coil. Direct expansion may be used instead of chilled water.

6.5.4 Air Washer with Mixed Air and Refrigeration

Mixed air can be used rather than 100% outside air if the control system is altered as shown in Figures 6–32 and 6–33. As long as the outside air wet bulb temperature is below the coil leaving-air set

Figure 6-31 Psychrometric chart for Figure 6-29 (summer).

Figure 6-32   Humidity control; sprayed cooling coil with mixed air.

point of T1, the desired condition can be maintained by adjusting the mixing dampers to obtain a mixture which falls on the coil leaving-air wet bulb line, and utilizing the evaporative cooling effect of the sprayed coil. When the outside air damper is fully open and the coil leaving-air temperature increases above the set point of T1, then refrigeration must be used, either chilled water as shown, or direct expansion. Again, reheat is required for space temperature control.

6.5.5   Steam Humidifiers

Steam humidifiers are often used because of their simplicity. A piping manifold with small orifices is provided in the air duct or plenum (Figure 6–34). The steam supply valve is controlled by a space or duct humidistat. Avoid the use of steam which has been treated with toxic chemicals. Almost any humidity ratio, up to saturation, can be obtained in the supply air stream.

Figure 6-33 Psychrometric chart for Figure 6-32.

Figure 6-34 Steam grid humidifier.

6.5.6 Pan Humidifiers

Evaporative pan humidifiers vary in capability and controllability. The old-fashioned evaporator pan found in many residential warm-air heating systems is not very efficient, and lacks any kind of control. The addition of a heater, of immersion or radiant type, improves efficiency and allows a limited amount of control. Even with heaters, however, this type of humidifier will not usually provide space relative humidity of over 40%.

6.5.7 Atomizer Humidifiers

Slinger or atomizer humidifiers may be controlled on-off by a humidistat or manually. Again, space relative humidity will seldom exceed 40%. During "off" periods the lime deposits left from the water evaporation may be entrained in the air stream as dust.

6.6 DEHUMIDIFIERS

As noted above, a sprayed cooling coil may serve as a dehumidifier if the entering air is sufficiently humid. However, to maintain very low humidities at normal air-conditioning temperatures it is necessary to employ chemical dehumidifiers or low-temperature refrigeration.

6.6.1 Chemical Dehumidifiers

Chemical dehumidifiers use a chemical adsorbent, usually with provision for continuous regeneration (drying) to avoid system shutdown. One form of dehumidifier (Figure 6-35) uses a wheel containing silica gel which revolves first through the conditioned air stream, absorbing moisture, and then through a regenerative air stream

Figure 6-35   Dehumidifier; chemical adsorbent.

of heated outside air which dries the gel. In the process, a great deal of heat is transferred to the conditioned air and recooling is necessary. The dehumidification efficiency is largely a function of the regenerative air stream temperature. Figure 6–36 shows the control system The space humidistat controls the heating coil in the regenerative air stream. Final control of space temperature is accomplished by the room thermostat and the recooling coil. Very low humidities may be obtained in this way.

Another type of system uses a water-absorbing liquid chemical solution which is sprayed over a cooling coil in the conditioned air stream, thus absorbing moisture from the air. A portion of the solution is continuously pumped to a regenerator where it is sprayed over heating coils and gives up moisture to a scavenger air stream which carries the moisture outside. Control of the specific humidity of the leaving conditioned air is accomplished by controlling the temperature of the solution. Since this is a patented, proprietary system, the control system recommended by the manufacturer* should be used.

* Ross Engineering Division of Midland-Ross Corp.

Figure 6-36   Control of chemical dehumidifier.

Figure 6-37   Dehumidifying with low-temperature cooling coil.

6.6.2   Dehumidifying by Refrigeration

Low-temperature cooling coils may also be used to reduce humidity to low values. Since coil surface temperatures will be below freezing, with ice formation resulting, special DX coils with wide fin spacing must be used, and provision must be made for defrosting by hot gas, electric heat or warm air. This approach tends to be inefficient at very low humidities and requires intermittent shutdown for defrosting, or parallel coils so that one may operate while the other is being defrosted. Reheat is necessary for control of space temperature. Since space humidity is largely a function of coil temperature, fairly good control may be achieved through humidistat control of a variable back-pressure valve (Figure 6–37). The selective relay allows the room thermostat to operate the cooling coil at minimum capacity when the humidistat is satisfied.

6.7   STATIC PRESSURE CONTROL

Static pressure controls are used to provide a positive or negative pressure in a space with respect to its surroundings. For example, a clean room will be positive to prevent infiltration of dust, while a chemical laboratory or plating shop will be negative to prevent exfiltration of fumes. Many nuclear processes require careful space segregation with fairly high pressure differentials. For most ordinary spaces it is impractical and unnecessary to design for pressure differentials greater than 0.1 in. water gage. Swinging doors are difficult to open and/or close even at this small pressure, and special sealing methods are necessary to maintain higher pressures.

Figure 6-38   Outside air; static pressure control.

6.7.1   Single Room

A pressure-controlled space may be served by its own air conditioning unit. Outside air may have minimum or economy cycle control by temperature, or may be simply 100%. Relief air is usually provided by power exhaust, although positive pressures in the space allow "gravity" relief. In either case, the static pressure controller controls the relief damper, and if 100% outside air is used, may also control the intake damper. Such a control system is shown in Figure 6–38. (See also Figure 6–7.) The reversing relay is needed so that both dampers can be normally closed.

6.7.2   Multiple Rooms

If several rooms are to be served by a single air conditioning unit then variable-volume dampers on supply and relief to each room will be operated by the static pressure controller for that room. Individual exhaust fans, or one large central exhaust may be used. Such a system is discussed in section 9.9.

6.8   ELECTRIC HEAT

Electric heaters may be controlled on the same basic cycles as other heating devices—two-position, timed two-position and proportional. Because of the use of electricity as the energy source, certain special considerations are necessary. Any electric heater must be provided with a high-limit control. Some codes require both automatic reset and manual reset high limits. Forced-air heaters should have airflow switches to prevent the heater from operating when the fan is off.

Figure 6-39   Electric heater; piloted control, two-position.

6.8.1   Two-Position Control

Two-position control of small-capacity heaters may be provided by a heavy-duty line voltage thermostat. More common, however, is the piloted system (Figure 6–39) with the thermostat operating a contactor. For large heaters it is common to use a multistage thermostat or sequencing switches with several contactors, each controlling current flow to a section of the heating coil.

6.8.2   Proportional Control

True proportional control may be obtained through the use of a saturable core reactor or variable autotransformer (Figure 6–40). A very small change in the DC control current can cause a large change in the load current. Since the controller must handle the full-load current it may become physically very large. Efficiencies are poor at part load.

Figure 6-40   Electric heater; saturable core reactor control.

Figure 6-41   Electric heater, time-proportional control.

6.8.3   Timed Two-Position Control

Timed two-position control can be achieved by means of a timer and the same contactors used for ordinary two-position control (Figure 6–41). The timer may be mechanical or electronic and provides a fixed time base (usually adjustable) of from one-half minute to five minutes. The percentage of on-time varies in accordance with the demand sensed by the room thermostat. The contactors will cycle on and off once in each time base period but the length of on-time will be greater if the room temperature is below the thermostat setting. However, rapid cycling, even with mercury switch contactors, may lead to maintenance problems. The preferred method is to use solid-state controllers: SCR's (silicon controlled rectifiers) or Triacs. These devices can provide extremely rapid cycling rates, so that control is, in effect, proportional (Figure 6–42). The solid-state controller has an electronic timer which provides an extremely short time base. The load current is handled directly by the controller through semiconductor

Figure 6-42   Electric heater; solid-state controller.

Figure 6-43   Electric heater; Vernier control system.

switching devices. The thermostat demand will vary the percentage of on-time. Power regulation may be accomplished by phase control or burst control. The SCR may be piloted by any type of proportioning thermostat, to vary the cycle time rate. The "Vernier control" system (Figure 6–43) combines sequence control with solid-state control to obtain near-proportional control. It is especially economical if the heater is large. The coil is divided into several small sections with one controlled by the solid-state unit and the others by sequencing step controller. As the space temperature decreases below the proportional thermostat setting the solid-state unit modulates its section of the coil from off to 100% on. If heating demand continues to increase the first section of sequence controlled portion is turned on and the power to the solid-state section may be cut back. This continues until all sections are on, and, of course, the opposite sequence occurs on a decrease in heating demand. The modulated section should have 25% to 50% greater capacity than the sequenced sections to prevent short-cycling at changeover points.

6.9   GAS-FIRED HEATERS

Direct gas-fired heaters are, generally, packaged equipment complete with controls. Usually the complete package, including controls, is approved by The American Gas Association (AGA). Any change in the system, however minor, will void this approval. There are still a number of options among the approved packages.

Figure 6-44   Gas heater; two-position control.

6.9.1   Two-Position Control

Two-position controls are most common (Figure 6–44). A variation of this is multistage two-position, with the gas burners sectionalized.

6.9.2   Proportional Control

Modulating gas valves are available. AGA approval for small systems is limited to self-contained types, with the sensing bulb mounted in the discharge air stream. Minimum capacity is about 30% of total capacity since the gas flame becomes unstable at lower rates (Figure 6–45). At this lower limit the valve becomes a two-position type.

6.9.3   Safety Controls

Any gas control system must include such safety controls as high-temperature limits and pilot flame proving devices. Forced-draft heating boiler controls may also include timed pre-purge and post-purge cycles to prevent gas accumulation and explosion. These specialized cycles are usually programmed by special-purpose electronic controls.

Figure 6-45   Gas heater; modulating control.

6.10   OIL-FIRED HEATERS

Oil burning heaters, like gas-fired units, generally come complete with a control package with UL approval. Two-position or modulating controls are available, depending on the type and size of the equipment. This is a highly specialized area. Residential and small commercial control systems will be discussed in Chapter 7.

6.11   REFRIGERATION EQUIPMENT

Central refrigeration equipment may include compressors, condensers and chillers. Compressors may be reciprocating, positive displacement (screw-type) or centrifugal. Often, the compressor is part of a package water chiller as described below. Reciprocating compressors, up to about 50 tons capacity, may be used directly with DX coils. Accurate control of DX systems above 25 or 30 tons is difficult and it is unusual to find such systems above 50 or 60 tons. Larger reciprocating machines, positive displacement units and centrifugal compressors will be part of a package chiller system, complete with factory installed controls. Absorption machines also come as part of a package chiller except for small residential units. Condensers may be air-cooled, water-cooled or evaporative, and are sometimes part of the chiller package.

6.11.1   Reciprocating Compressors

Reciprocating compressors, except in very small sizes, have multistage capacity control. This is generally achieved by "loading" and "unloading" cylinders under control of a suction pressure or chilled water temperature controller, by raising the suction valve off its seat (Figure 6–46). Unloading devices may be mechanical or elec-

Figure 6-46   Refrigeration compressor unloader.

Figure 6-47   Refrigeration compressor; "pump-down" control.

trical and are always a part of the compressor package. The number of steps is determined by the manufacturer as a function of size, number of cylinders and machine design.

Starting and stopping the compressor may be done directly by the room or chilled water thermostat, but is more often done on a "pump-down" cycle. On a rise in temperature of the controlled medium the thermostat opens a solenoid valve in the refrigerant liquid line to the chiller or cooling coil. Refrigerant flow raises the suction pressure. The low-pressure switch closes and starts the compressor. When the thermostat is satisfied and closes the solenoid valve, the lowering of the suction pressure as refrigerant is pumped out of the evaporator opens the low-pressure switch, stopping the compressor. Figure 6–47 shows a typical electric control system for a medium-sized reciprocating compressor, with safety and operating controls.

6.11.2   Centrifugal, Positive-Displacement and Absorption Chillers

Since centrifugal and absorption machines always occur as part of a chiller package, the complete control system will be furnished by the manufacturer. These systems usually include fully modulating capacity controls with a low limit of 20% to 30% of maximum capacity, and provide elaborate safety and interlock controls to protect the equipment. A typical control system for a centrifugal package is shown in Figure 6–48. The chilled water and condensing water pumps are started manually (or the condensing water pump may be interlocked to start when the chilled water pump is started). When the chilled water thermostat calls for cooling the compressor will start, provided that flow switches and safety switches are closed. The thermostat will then modulate the inlet vane capacity controller, with the

Figure 6-48   Centrifugal water chiller control.

capacity limiting controller acting as a maximum capacity limiting device. When load falls below about 20% of machine capacity the thermostat will stop the compressor. A time-delay relay, not shown, is
usually provided to prevent restarting the compressor at less than 30
min intervals. This is necessary to prevent damage to the reduced-
voltage starter.

6.11.3   Air-Cooled Condensers

Air-cooled condensers use ambient air forced across a condensing coil to cool and liquify the compressed refrigerant gas. The
condenser fan is controlled to start automatically whenever the compressor runs. Condensing pressures at high ambient temperatures will
be higher than with water-cooled or evaporative condensers, since
these depend on wet bulb temperatures. But, at low ambient temperatures, this condensing pressure may fall so low as to cause operating problems in the refrigerant system. Any air-cooled system which
must operate at low ambient temperatures must be provided with
"head-pressure" control.

A simple head-pressure control system (Figure 6–49) uses modulating dampers to reduce airflow, finally stopping the fan when the
damper is nearly closed. Alternatively, a variable-speed fan may be
used.

A more elaborate and effective system uses the "flooding" principle,
in which a throttling valve slows down the flow of liquid refrigerant,
causing the condenser coil to be partially or wholly filled with liquid,
thus reducing its capacity. There are several "flooding control" systems, mostly patented by manufacturers.

Figure 6-49   Air-cooled condenser.

6.11.4   Water-Cooled Condensers

Water-cooled condensers use water from a cooling tower or other source to cool and liquify the refrigerant gas. Condensing pressure can be closely controlled by modulating water flow or controlling supply water temperature by mixing condensing return water with supply water. If a cooling tower is used, water temperature may be controlled by cycling the tower fan or bypassing some of the tower flow (Figure 6–50). With the temperature of the condensing water supply (CWS) above the thermostat set point the flow valve V1 is open, bypass valve V2 is closed and the fan is running. As the CWS temperature decreases below the set point the fan is turned off. On a further decrease valve V1 closes and V2 opens. As the CWS temperature increases the reverse cycle takes place. Absorption chillers require close control of condensing water temperature, with "mixing" most often used (Figure 6-51). The CWS thermostat modulates the mixing valve to maintain a constant supply water temperature. When the valve is in full bypass condition, the cooling tower fan is stopped. In order for the bypass valve to function properly it is necessary to have several feet of gravity head above the valve to the tower. If a good static head is not available a diverting valve may be used. (Figure 6–52).

Figure   6-50   Cooling   tower;   bypass   valve control.

Figure 6-51   Cooling tower; three-way valve control.

6.11.5   Evaporative Condensers

Evaporative condensers take advantage of wet bulb temperatures by spraying water over the refrigerant condensing coil. Air is forced or drawn across the coil and the resulting adiabatic saturation process provides efficient condensing. Control of condensing temperature is achieved by varying the airflow (Figure 6-53). The head-pressure controller modulates the outside air damper to provide constant head pressure. When the damper is nearly closed the fan is stopped.

6.11.6   Water Chillers

Chillers may not be part of a package. When installed as a separate piece of equipment two arrangements are possible. Flooded systems use a surge tank with low- or high-pressure float to control

Figure 6-52   Cooling tower; three-way diverting valve.

Figure 6-53   Evaporative condenser control.

refrigerant feed. Direct-expansion systems use a thermostatic expansion valve. In either case refrigerant flow will be started or stopped by a solenoid valve in the refrigerant liquid line, controlled by a chilled water thermostat. Interlocks to improve chilled water flow and prevent freezeup are required. Figure 6-54 shows a direct-expansion chiller with water flow and low water temperature interlocks. The operating thermostat, T1, in the chilled water return will open the solenoid valve in the refrigerant liquid line, providing that the water is flowing and not near freezing.

6.11.7   Cooling Towers

Cooling towers which will operate only during the cooling season are usually provided with fan control only. A condensing water supply thermostat will start or stop the fan as the temperature rises or falls. On larger towers two-speed fans may be used, with two- or three-stage thermostatic control.

Figure 6-54   Water chiller control.

Figure 6-55   Cooling tower; winter operation.

Towers which are used all year need more extensive control systems, including bypass valves, and heating to prevent freezeup. Figure 6-55 shows a year-round system with two-speed fan control, bypass and heating. On small or medium-sized systems (to about 200 tons) towers can sometimes be installed indoors so that modulating dampers to vary airflow will provide adequate control (Figure 6-56). When the damper is fully closed the fan will be stopped. An outdoor tower with an indoor sump may also be used.

6.12   ELECTRICAL INTERLOCKS

As has been indicated in several of the preceding examples, at some point it will be necessary for the temperature and pressure

Figure 6-56   Cooling tower; indoor arrangement.

controls to interface with electric motor controls. Small motors may sometimes be operated directly by electric controllers. If the motors are large, or of a different voltage than the control circuits, or if pneumatic or electronic controls are used, then relays are required. This subject is discussed more fully in Chapter 8.

6.13   LOCATION OF SENSORS

The proper location for the sensor can best be determined by asking specifically what is to be controlled. If it is room temperature, then the location should be such that the sensor reads an average room temperature, with a minimum of exposure to supply air, drafts or radiant effects from windows or equipment. Near, or even in, return air openings is recommended.

In a large work area, you may want the best control at a particular work station. So, mount the sensor there. Sensors have been mounted on movable frames with long cables, and relocated as the critical work station changed.

The main thing to avoid is "side effects" which may prevent the sensor from seeing conditions correctly. For temperature sensors these may be radiation (cold or hot), drafts, lack of adequate air circulation or convective heat transfer through the mounting (as on an outside wall).

In systems with several rooms on a single zone, it is essential to select an "average" room for the sensor location. Conference rooms or rooms with large load variations should be on separate zones, but, if this is not possible, do not let them be the sensing point for the zone. Given a choice between large and small rooms, select the larger space (unless the smaller space is used by the manager's executive secretary!). All rooms on a single zone should have comparable outside exposures with a common orientation (north, south, east or west, but not a mixture of these).

6.14   SUMMARY

This chapter has discussed elementary control systems or, more correctly, subsystems. Any larger system is built up from combinations of these elements. In Chapters 7 and 9 it will be shown how this is done. It is hoped you will be able to recognize these smaller elements in the larger systems.

7 Complete Control Systems

7.1 INTRODUCTION

In the previous chapter we discussed elementary control systems. These are, in general, the pieces and parts which can be fitted together to produce specific applications solutions.

This chapter will discuss and detail many of these applications as they have been or could be solved. The discussions proceed from the simple single-zone concept to more elaborate arrangements. Some highly complex systems will be dealt with in Chapter 9.

7.2 SINGLE-ZONE SYSTEMS

Single-zone systems have one basic parameter: a single zone temperature controller. This controller may serve one or more air handling units and may be a simple room thermostat, a discharge thermostat or a complex averaging or reset controller.

7.2.1 Single Air Handling Unit

Consider first a simple and common type of system consisting of a single air handling unit (AHU) with a room thermostat as the

Figure 7-1   Single-zone AHU; minimum outside air.

basic controller. For year-round air conditioning this AHU will have heating and cooling coils and sufficient outside air for ventilation. If the coils use hot and chilled water then the simple system might look as shown in Figure 7–1. When the supply fan is started (manually) the minimum outside air damper opens and the temperature control circuit is energized. The room thermostat then opens and closes the hot and chilled water valves in sequence to provide heating or cooling as required. Notice that only one of these valves may be open at one time. The fire safety switch will stop the fan if a high (about 125° F) return air temperature is sensed. A smoke detector may be used instead. The diagram shows the heating coil preceding the cooling coil. The primary reason for this is to minimize coil-freezing possibilities. There is considerable disagreement as to which sequence (cold-hot or hot-cold) should be used. Generally we favor the hot-cold approach except when maximum humidity control is required. Then the "reheat" position is necessary for the hot coil. (See Figure 7–6.) The system just described will have a rather wide operating differential even with a very small control differential. For example, if the thermostat senses a rise in room temperature and opens the chilled water valve it will take some time for the water to flow into the coil, lower its surface temperature and cool the air passing over it. More time is required for the air to travel to the room, mix with room air and lower the temperature at the thermostat location. The thermostat then closes the valve, but by this time the coil is full of cold water, which causes a continuing decrease in room temperature. The same sequence occurs on heating except that the overrun is usually greater due to the greater initial temperature difference between the hot water and the air.

Because this situation exists the system is frequently modified by

Figure 7-2 Single-zone AHU; minimum outside air, discharge thermostat.

adding a thermostat in the air stream leaving the AHU (Figure 7–2). The discharge thermostat may be used as a high-limit controller. If the discharge temperature exceeds the high-limit setting the discharge thermostat overrides the room thermostat and closes the hot water valve. This operation tends to limit the operating differential. A low-limit controller may also be used. The discharge thermostat may also be used as the basic "submaster" controller for the hot and chilled water valves, with set point being reset by the "master" room thermostat. This means that when the room thermostat senses the need for cooling, instead of controlling the valves directly, it causes a decrease in the set point of the discharge thermostat. And, when heating is required, the reverse is true. Since the discharge thermostat senses system response very quickly, system differential is decreased. The "reset schedule" may vary over a 30° to 40° range of the discharge temperature.

These "improvements" are not really needed for small zones, which respond quickly, but are highly recommended where a large area is served as a single zone.

7.2.2 Single-Zone Unit, Economy Cycle Outside Air

To this point in the chapter only fixed outside air has been considered. But, as noted in Chapter 6, a great many systems use the "economy cycle" control of outside air. If economy cycle control is added, with a separate relief exhauster, the system shown in Figure 7–3 is obtained. This is the system most commonly used today. Here, the fan is started manually, with interlocks to start the relief fan and allow control operation when the fan is running. The outside, return

Figure 7-3   Single-zone AHU; economy cycle outside air, discharge controller.

and relief dampers modulate in response to the mixed-air controllers. The discharge thermostat controls the hot and chilled water valves in sequence. The room thermostat resets the discharge control point. Note that the diagram shows a single controller with two sensors. This is the way many manufacturers now provide this control, but it may also be two separate units in a master-submaster arrangement.

Figure 7-4   Single-zone AHU; economy cycle, discharge controller, static pressure control.

7.2.3 Single-Zone Unit, Static Pressure Control of Outside Air

By the simple addition of a static pressure controller to the above cycle it is possible to maintain a positive or negative pressure in the room (Figure 7–4). This controller operates the relief damper independently of the return and outside air dampers, thereby maintaining the set pressure. A throttling device in the control output prevents overshoot or hunting due to sudden changes such as opening a door.

It should be noted again here that room static pressures greater than $\pm$ 0.10 in. of water are difficult to maintain unless all opening to other spaces are weatherstripped. Such pressure differentials also cause difficulty in opening or closing doors and related problems. All pressurization requirements should be carefully studied, and the minimum suitable value should be used.

7.2.4 Return-Relief Fans

Large air handling systems, or those with lengthy return air ducts, should be provided with return-relief fans (Figure 7–5). Return air quantity is equal to supply air quantity less any fixed exhaust and exfiltration requirements. A regular economy cycle then modulates outside air, return air and relief dampers, and additional relief fans are not required.

Figure 7-5   Single-zone AHU with return-relief fan.

Figure 7-6   Single-zone AHU; maximum humidity control by reheat.

7.2.5   Single-Zone Humidity Control

As suggested earlier, control of maximum humidity can be provided by placing the cooling coil ahead of the heating coil in the airflow sequence. Then a space humidistat is added to the control sequence (Figure 7–6). Now, so long as room humidity is below the humidistat setting, the system operates as in Figure 7–5. If the humidity goes above the set point, the humidistat, through the selector relay, takes over control of the cooling coil, and calls for additional cooling, which provides dehumidification. If the room then becomes too cool, the thermostat will call for "reheating." The psychrometric chart in Figure 7–7 shows the cycle graphically.

Year-round humidity control within fixed high and low limits can be accomplished in several ways. One frequently used method is the "sprayed-coil dehumidifier" described in chapter 6. A complete air handling system, using 100% outside air, preheat, sprayed-coil dehumidifier and reheat, is shown in Figure 7–8. The matching psy-

Figure 7-7   Psychrometric chart for Figure 7-6.

Figure 7-8   Single-zone AHU; humidity control with sprayed coil, 100% ouside air.

chrometric chart cycle is shown in Figure 7–9. The thermostat in the sprayed-coil leaving air is set for a dry bulb temperature which assumes saturated air leaving the coil. This setting is below the room design dew point by a specific humidity difference corresponding to the room latent load. This "dew-point thermostat" controls the preheat coil to provide sufficient preheat to allow the sprayed coil to act as an evaporative cooler. If the outside air wet bulb temperature is above the coil leaving temperature then the leaving-air thermostat will call for refrigeration (DX or chilled water) to achieve the set-point condition. The room thermostat controls the reheat coil to maintain room dry bulb temperature requirements. This system will easily maintain relative humidity between 40% and 60% within ± 5% of re-

Figure 7-9   Psychrometric chart for Figure 7-8.

Figure 7-10   Single-zone AHU; humidity control with sprayed coil and mixed air.

quirements or closer if room dry bulb temperature is maintained. To change the room humidity control point it is necessary to raise or lower the dew-point thermostat setting. Thus, humidity control is limited in the lower range by the refrigeration capacity of the system and in the upper range by the decreasing dry bulb temperature difference between supply air and room, since there is a limit to the total air that can be supplied without causing drafty and/or noisy conditions.

A similar control cycle can be supplied if the system requires less than 100% outside air (Figure 7–10 and psychrometric chart 7–11). The preheat coil is omitted. The dew-point thermostat controls outside air, return and relief dampers as follows: If the coil leaving air temperature decreases below control set point, the outside air and

Figure 7-11   Psychrometric chart for Figure 7-10.

relief dampers modulate toward the closed position and the return air damper modulates toward the open position. What is really being sensed, of course, is the mixed-air dry bulb-wet bulb temperature condition necessary to achieve the proper leaving-air dew-point temperature. As the leaving air temperature increase the opposite sequence takes place, until the outside air reaches 100%. Now any further increase in the outside air dry or wet bulb temperature will cause the controller to call for refrigeration. At some outdoor dry bulb temperature above this point, therefore, it is necessary to close the outside air damper to a minimum position (and position return and relief air dampers accordingly). This is done by an outside air high-limit thermostat, as described in Chapter 6. The room thermostat controls the reheat coil to provide room temperature control. This cycle is subject to the limitations described in the preceding paragraph.

If higher humidity or a simpler system is desired, humidifiers may be used, as described in Chapter 6. The humidifier is inserted in the duct or plenum downstream of the cooling coil and controlled by a room or duct humidistat. A few words of advice regarding humidifiers in general would not be amiss here.

Any evaporative system, such as a pan, or sprayed coil, generates a severe maintenance problem, due to deposition of solids from the evaporated water. It has been found, through experience, that a demineralizer on the makeup water will often pay for itself by saving the cost of cleaning up coils and pans.

Steam humidifiers should not be used where the steam is treated. Many steam system additives are toxic, besides causing odors or unsightly deposits. Where a central steam or high-temperature water supply is provided and you wish to use a steam humidifier, a small steam-to-steam or hot-water-to-steam generator can be used to provide clean steam.

7.3 MULTIZONE AIR HANDLING SYSTEMS

In multizone systems there are two related and dependent control problems: control of air temperatures in hot and cold plenums, and control of temperatures in the zones. It is very easy to lose some control if the hot plenum temperature is too high, and the mixing dampers have a large leakage factor (over 10%). Modern, tight-closing dampers help to overcome the problem but it is still highly desirable to limit the hot plenum temperature to the lowest value required for

control in the coldest zone. On the other hand, cold plenum temperature will not vary greatly even if uncontrolled, so that frequently the
control valve is omitted on chilled water coils. For optimizing of large
systems a chilled water control valve is required.

An additional source of difficulty occurs at light-load conditions
with modulating controls. Then, the temperature gradient across the
face of the coil, from one end to the other, may be 5° or 10° or more.
Obviously this causes lack of adequate control, especially for zones
which feed off the plenum near the ends of the coils. This can be
solved by the use of circulating pumps with water coils, by the use of
"distributing-tube" steam coils or by feeding both ends of long coils.
With direct-expansion coils a two- or three-stage sequence, in which
one or more rows *in the direction of airflow* constitute a stage, has
been used successfully.

7.3.1   Multizone Control

Figure 7–12 shows a typical multizone control system. The
return-relief fan and outside air controls have been described before.
Usually the mixed-air high- and low-limit settings are such that no
refrigeration is required in cold weather and no heating is required in
hot weather. The bypassed mixed-air temperature is adequate for
zone control. The cooling coil is uncontrolled, but chilled water may
be shut off manually in cold weather. The hot plenum temperature is

Figure 7-12   Multizone system; hot plenum reset.

Figure 7-13 Multizone system; hot plenum control from zone with greatest demand.

controlled by a plenum thermostat with reset from outdoor temperature. An estimate of the hot plenum temperature required for adequate heating of a central zone at various outdoor temperatures is the basis for the reset schedule. Each zone thermostat controls the mixing dampers for its zone to maintain design zone conditions.

Figure 7–13 is the same as Figure 7–12 except for the hot plenum control. Here, in order to keep the hot plenum temperature at the lowest possible value it is necessary to use a relay or series of relays from the zone thermostats, so that control of the heating coil is given to the thermostat with the greatest demand for heating.

7.3.2 Multizone Control, 100% Outside Air

Multizone units may need 100% outside air. In this case, preheat is required, usually to 50° or 55°, which provides a cold plenum temperature adequate for cooling interior zones. Figure 7–14 shows such a system.

7.3.3 Multizone with Humidity Control

A multizone system may also be used for humidity control. Figure 7–15 shows a suitable arrangement. The outside air-return air mixing dampers and sprayed cooling coil are controlled as described

Figure 7-14   Multizone system; 100% outside air.

in section 7.2.5 and Figure 7–10. The heating coil is controlled by the zone with the greatest demand, as described above. The zone thermostat then controls the zone mixing dampers to maintain design zone temperature. The psychrometric chart is similar to Figure 7–11. Relative humidity control will be very accurate (±5% or less) if all zones are at essentially the same temperature.

7.4   DUAL-DUCT SYSTEMS

Dual-duct systems, whether high- or low-velocity, are simply multizone systems with the plenums extended and the zone mixing dampers located near the zone. Zones are generally smaller and response is faster. High-velocity, high-pressure dual-duct systems also pose problems of large static pressure differentials between hot and cold ducts, which may affect the performance of the mixing dampers. This usually occurs during periods of maximum heating or cooling loads, when one side of the system is handling a large air quantity compared to the other side. This results in higher friction losses in the high-flow side and high static pressure differential at the mixing box.

Figure 7-15 Multizone system; sprayed coil for humidity control.

7.4.1   Dual-Duct Static Pressure Control

Static pressure differentials may be minimized by static pressure control dampers in the main ducts, near the air handling unit. The sensor measures static pressure near the end of the duct and modulates the damper between full open and some minimum closure position to maintain the desired end-of-main pressure. This must, of course, be set high enough to properly serve the most remote mixing box. When both ducts have such controls the end-of-main differential should be small. It should be noted that many systems operate satisfactorily without static pressure controls, and the present-day mixing boxes work very efficiently even with large pressure differentials.

7.4.2   Mixing Units

Mixing boxes for high-pressure systems should be selected with a high enough pressure drop for good control, without being noisy. Between one and two inches water pressure drop seems to be optimum. Most mechanical constant-volume controllers do not give satisfactory operation below about 0.75 in. of water pressure drop. Many older systems used two-motor constant-volume controllers and these will operate satisfactorily at pressure drops as low as 0.25 in. of water. Most systems today use mechanical constant-volume controllers. These take various forms, but are all basically spring-loaded dampers which open on a decrease in upstream pressure and modulate toward a minimum closure position as pressure increases.

Figure 7–16 shows a one-motor, mechanical constant-volume mixing box control. The dampers are operated in parallel by the motor, so

Figure 7-16   Mixing box; mechanical constant-volume.

Figure 7-17   Mixing box; two-motor, constant-volume.

that as one closes the other opens. Pressure changes which occur during the cycle, as a result of pressure differential between the hot and cold ducts, are compensated for by the mechanical constant-volume controller. The room thermostat positions the damper motor to provide more hot or cold air as required.

Figure 7–17 shows a two-motor constant-volume controlled mixing box. The room thermostat modulates the hot damper motor toward open or closed position as required to maintain room temperature. The constant-volume controller senses the change in pressure drop through the mixing box, caused by the hot damper change, and positions the cold damper to compensate and thus maintain a constant flow through the box. The relay allows the volume controller to override the room thermostat and partially open the hot air valve when the cold air valve is full open but total volume is too low. This happens occasionally on high cooling demand.

This type of system is still used for zones larger than can be handled by a standard mechanical constant-volume mixing box.

Low-pressure mixing box systems are also used, though not widely, since they are usually limited in size, due to economics and space available for duct work. One-motor boxes without constant-volume control are normally used, since available pressure drops are not sufficient for constant-volume control.

7.4.3   Dual-Duct Air Handling Unit Controls

Controls for the air handling units in dual-duct systems are identical with those used for multizone units. Outdoor reset of hot plenum temperature is commonly used.

7.5   VARIABLE-VOLUME SYSTEMS

Variable-volume systems provide multizone control with only a single duct. The supply air is maintained at a constant temperature which varies with the season. The individual zone thermostat varies the air supply quantity to the zone to maintain the desired temperature condition. Minimum supply air quantity is usually not less than 50% of design airflow to provide sufficient ventilation. Obviously, all zones on a system must have similar heating and cooling requirements. It is not possible to provide simultaneous heating and cooling with this system, as it is with a multizone or dual-duct system.

Figure 7–18 shows a variable-volume system. The supply air temperature is maintained near cooling design set point so long as the system is on "cooling." Zone thermostats vary airflow to each zone as required. Changeover to "heating" changes the set point of the supply air thermostat (T1) and reverses the action of all the zone thermostats. For example, on "cooling," the zone thermostat modulates the normally open zone damper toward the minimum closure position as zone temperature decreases and opens the damper as zone temperature increases. On "heating," the opposite sequence takes place. Reversal of the controller operation is accomplished in various ways, depending on whether it is pneumatic, electric or electronic. Changeover from heating to cooling may be accomplished manually, as shown, or automatically by means of an outdoor thermostat or critical-zone thermostat.

It is most important to remember that a minimum airflow must be maintained and all zones must have similar heating and cooling requirements. (See also Figure 9-8.)

7.6   REHEAT SYSTEMS

Reheat systems also provide multizone control with a single duct (Figure 7–19). The air supply temperature is essentially constant at a value suitable for cooling all year round. Reheat coils for each zone are controlled by zone thermostats to satisfy zone temperature requirements. This provides more flexibility than variable volume since that system can now provide simultaneous heating and cooling. It also costs somewhat more to install and operate. The heating coil in the AHU may be omitted if minimum mixed-air temperature is above 50°.

Figure 7-18 Variable-volume system.

7.6.1 Reheat with Variable Volume

Some of this operating cost disadvantage may be overcome by combining reheat with variable volume (Figure 7–20). Air is supplied from the control unit at a constant temperature suitable for cool-

Figure 7-19 Reheat system.

ing, say 55°. When the zone air temperature is above the thermostat setting the volume damper is fully open and the reheat is off. As the zone temperature approaches the set point, the damper modulates toward some minimum position (usually not less than 50%). A further drop in zone temperature will cause the reheat coil to start heating.

Figure 7-20 Reheat plus variable-volume.

Figure 7-21   Humidity control; sprayed cooling coil with zone reheat.

This system is more economical than simple reheat because less re-heating is required, yet it provides simultaneous heating and cooling in adjoining zones.

7.6.2   Reheat for Humidity Control

Zone reheat may also be used for controlling humidity when the dew point of the main supply air is held constant (Figure 7–21). This system is similar to the single-zone system discussed in section 7.2.5 and Figure 7–8 except that now there may be several zones, each with individual temperature control. Since the supply air dew point is the same for all zones, the actual relative humidity in a zone is determined by its dry bulb temperature and internal latent load. It follows that these must be similar from zone to zone if rela-tive humidities are to be uniform.

The supply air dew point may be controlled by using a sprayed coil as shown above, or by means of humidifiers, as previously discussed.

7.7   HEAT RECLAIM

In general terms, a heat reclaim system utilizes heat (or cold) that would otherwise be wasted. Heat reclaim systems thus may include heat pumps, but are by no means limited to them. A typical heat reclaim system may use the excess heat generated in the interior

Figure 7-22 Heat reclaim; heat pump only.

zones of a building to heat the exterior zones in winter, while at the same time using the cooling effect of the exterior zones to cool the interior. Or, where large quantities of exhaust air are required, with equivalent outside air makeup, the waste heat from the exhaust may be used to preheat (or precool) the makeup air.

7.7.1 Heat Reclaim—Heat Pump

Reclaim systems with heat pumps take advantage of the fact that almost all buildings have interior areas with little or no exposure to outdoor conditions. Since these spaces do contain lights and people and other heat-producing devices, such as office machines and computers, they require cooling at all times. Or, looking at it another way, they have excess heat which must be removed to maintain comfort conditions. Since the same building will also have exterior zones which require heating in winter it is possible to use the interior-zone excess heat as a heat source for a heat pump which supplies heat to exterior zones. Figure 7–22 shows an elementary schematic of such a system. Chilled water from the heat pump cools the interior zone, the rejected heat is carried by the refrigerant to the condenser and the condenser water is used to heat the exterior zone. As can readily be seen, this elementary idea works only when there is a balance between interior cooling and exterior heating loads, so some additional provisions are necessary.

One common element is a "split" condenser (Figure 7–23). Now, that part of the interior heat which is not needed for exterior heating can be dissipated to a cooling tower or evaporative condenser. If a heat source, such as a boiler or electric heater, is added for those times when exterior losses exceed interior gains, a year-round system is obtained. It is desirable, but not essential, that separate air handlers be used for interior and exterior areas.

Figure 7-23 Heat reclaim; heat pump with auxiliary heat source and sink.

The interior-zone air handler, which could be any of the types previously discussed, provides cooling. The rejected heat, in the form of an increase in chilled water temperature, goes back to the heat pump, where chilled water supply temperature is the controlled variable. The rejected heat is conveyed by the refrigerant to the condenser water, which in turn is used as a heat source for the exterior-zone air handling unit heating coil. Water temperature leaving the condenser is controlled by a thermostat which operates as follows: When the water temperature is below the thermostat setting it calls on the auxiliary heat source for additional heating. As the water temperature increases the auxiliary source is shut down, and on a further increase, the other section of the split condenser is brought into play, dumping the excess heat to the cooling tower. In a small system this same chiller-heat pump could also supply cooling for the exterior zones, but in larger systems an additional chiller (not in a heat pump arrangement) will be provided.

This arrangement is used most often in high-rise office buildings or similar structures with interior areas which comprise a fairly large portion of the total floor area.

7.7.2   Heat Reclaim—Revolving Wheel

The revolving wheel heat reclaim system uses a large metal wheel which revolves slowly as exhaust air passes through one side and makeup air passes through the other. Heat is thus transferred by the turning wheel. The only control required is an interlock to start the wheel drive motor when the supply fan is operating. These devices

Figure 7-24   Heat reclaim; runaround coils.

are produced by several manufacturers and are very satisfactory where system geometry allows makeup and exhaust ducts to be contiguous.

7.7.3   Heat Reclaim—Runaround Coils

The "run around" heat reclaim system uses coils in exhaust and makeup air ducts, with piping connections and circulating pumps. While obviously more expensive than the revolving wheel, this system allows complete flexibility of system geometry and provides high efficiencies. Eight row coils will allow over 40% of the waste heat to be reclaimed in winter and about 25% of the waste "cooling" in summer. Where freezing air is encountered the system is usually filled with an ethylene glycol solution. Figure 7–24 shows such a system. The circulating pump is interlocked to run whenever the supply fan runs. No additional controls are required since it is necessary to operate at maximum capacity at all times.

7.8   FAN-COIL UNITS

Fan-coil units are widely used in hotels, motels, apartments and offices. Essentially, they are small single-zone air handling units with a fan, a filter and a coil which may be used for both hot or chilled water or may be split, with one row used for heating and two or three used for cooling. The former arrangement is more common since it is less expensive and requires fewer controls.

The control system may be arranged for two-pipe, three-pipe or four-pipe supply, each of which has certain advantages and disadvantages. When a single coil is used for both hot and chilled water, hot water temperatures may be much lower than with standard heating coils. This is because of the extra surface required for cooling at the normally low temperature difference between chilled water and leaving air. Hot water temperatures of 110° to 140° are typical.

7.8.1   Two-Pipe   System

Water is used, and hot or chilled water is supplied from a central plant in season. The room thermostat must be a summer-winter type—that is, it is direct-acting in winter, reverse-acting in summer when used with a normally open valve. Changeover from one to the other is accomplished by a general signal, such as changing main air pressure in a pneumatic system, or by sensing the change in supply water temperature. The thermostat may modulate the water flow valve or simply start and stop the fan while the water flow is uncontrolled. Figure 7–25 shows a common arrangement with manual multi-speed control of the fan and thermostatic control of the water flow to the coil. With chilled water being supplied and the room temperature above the reverse-acting thermostat setting, the normally open valve is open, allowing the unit to provide cooling. As the room temperature falls the valve closes. On a further fall in room temperature, no heat is provided since hot water is not available.

With hot water being supplied the room thermostat is direct-acting and the reverse cycle takes place. Now, of course, no cooling is possible. It is apparent that this arrangement can cause loss of control in mild weather, when heating is required in morning and evening but cooling is required during the day.

A central plant for a two-pipe system includes a boiler or other heat source, a chiller, circulating pumps and changeover controls. Changeover may be manual or automatic, but in either case involves prob-

Figure 7-25   Fan-coil unit; two-pipe system, manual fan control.

Figure 7-26 Two-pipe system; central equipment and control.

lems. The principal difficulty arises when changing from heating to cooling. If the water entering the chiller is too warm—and "too warm" may be mild temperatures such as 75° or 80°—the chiller becomes overloaded, with high suction pressure and possible compressor damage, or, at least, a safety shutdown. Therefore it is necessary to provide a bypass around the chiller and pass only part of the water through it until all the system water is cooled down. Obviously, this takes time, so spring and fall weather with warm days and chilly nights cause difficult and uneconomical operation. Water temperature reset is usually provided to minimize the changeover problem. Figure 7–26 shows a recommended control system. When the outdoor temperature is below about 60° the system is providing hot water with the supply temperature reset from about 140° at 0° outside to perhaps 75° at 60° outside. At approximately 60° the system changes automatically from heating to cooling by means of four two-position valves (two three-way valves can also be used). A temperature-controlled three-way valve diverts part of the water around the chiller, gradually closing the bypass as the system cools down. Note that minimum flow through the chiller must be maintained at all times to prevent freezing. The chiller solenoid valve is controlled by the leaving chilled water thermostat.

Changeover from cooling to heating may cause thermal shock in the boiler, and a heating bypass valve or a heat exchanger is preferably used. Water temperatures below 120° may also cause flue-gas condensation and resultant corrosion in the boiler.

7.8.2   Three-Pipe System

Water is used, with separate hot and chilled water supply pipes and a common return pipe. Individual heating and cooling valves or a special three-pipe valve may be used. A standard direct-acting thermostat is required. Since it is possible, in season, to provide a choice of heating or cooling, this system is much more flexible than the two-pipe system. Mixing the return water, however, means that when both heating and cooling are being used the return water is at some intermediate temperature and a false load is placed on both chiller and boiler, with a resultant increase in operating cost.

Figure 7–27 shows a typical three-pipe fan-coil arrangement, using a special three-pipe valve. When the room temperature is below the thermostat setting the heating port of the three-pipe valve is open. As the room temperature increases the hot water port closes. There is then a "dead spot" over a small temperature range, during which both ports are closed. As the room temperature goes still higher the chilled water port opens. Chilled or hot water flows out to the common return main where it is mixed with chilled or hot water return from other units and conveyed back to the central plant.

A central plant for a three-pipe system includes a boiler or other heat source, a chiller, a circulating pump and temperature controls. Since the water flows through boiler and chiller are determined by the sums of the individual fan-coil usages, the system is self-balancing in this regard. Return water temperature is directly related to load, increasing when heating demand is increased and decreasing when cooling demand is increased. This seeming anomaly results from the fact that return hot water is hotter than return chilled water. Since the flow through the chiller will decrease as the return water temperature rises, there is little need for the elaborate temperature-flow controls discussed under two-pipe systems. However, a sudden increase in cooling load can cause problems. Also, low-temperature water going

Figure 7-27 Fan-coil unit; three-pipe system, manual fan control.

Figure 7-28   Three-pipe system; central equipment and control.

through the boiler can cause condensation of flue gases in the boiler, with resulting corrosion. For this reason, a heat exchanger is recommended for two- and three-pipe systems requiring hot water to be supplied at temperatures below 140° Figure 7–28 shows a typical three-pipe central plant control arrangement. The individual water supply thermostats control the heat exchanger and chiller capacities. A flow switch on the chiller will shut off the compressor if water flow falls below the minimum rate necessary to prevent freezing.

7.8.3   Four-Pipe System

Four-pipe systems provide complete segregation of the heating and cooling media, thus avoiding the problems of two- and three-pipe installations. The normal coil configuration is a split arrangement with one row used for heating and two or three used for cooling. Sometimes two separate coils are used. With this set-up virtually any heating and cooling source can be used, including steam, hot or chilled water, direct-expansion cooling and electric resistance heating.

Figure 7–29 shows a typical split-coil control arrangement. Here, steam is used for heating and chilled water for cooling. With the room temperature below the setting of direct-acting thermostat, the normally open steam valve is open and the normally closed chilled water valve is closed. As the room temperature increases the steam valve closes. The controls should be adjusted to provide a "dead spot"

Figure 7-29   Fan-coil unit; four-pipe system, split-coil.

where both valves are closed. As the room temperature rises still further, the chilled water valve opens. On a fall in temperature, the sequence is reversed.

It is also possible to use a single coil for either hot or chilled water, while keeping the piping systems segregated. This requires the use of two "three-pipe" valves.

Figure 7–30 shows such an arrangement. When the room temperature is below the setting of the thermostat the inlet and outlet valves are open to the hot water supply and return lines. As the temperature increases, the inlet valve heating port may modulate toward the closed position, but the outlet valve heating port will remain full open to minimize pressure drop. When the inlet valve is fully closed then the outlet valve will close, using two-position action. Again, there will be a period when both ports are closed and room temperature is at or near the thermostat setting. On an increase in room temperature above the setting, the inlet and outlet valve cooling ports will modulate toward open, or open using two-position sequence.

7.9   INDUCTION SYSTEMS

An induction unit depends for its operation upon a central source of temperature-controlled high-pressure air, which induces a secondary flow of room air across the unit coil. This coil is supplied with hot or chilled water through a two-, three- or four-pipe system. Thus it is necessary to control the primary air supply temperature, the

Figure 7-30   Fan-coil unit; four-pipe system, single-coil.

temperature of the hot or chilled water and the flow of this water. Since condensate drains are not usually provided on induction unit coils, the chilled water supply temperature must be kept at 55° or higher. Because a single coil is used for both heating and cooling, hot water supply temperatures can be comparatively low, in the order of 140° or less.

Therefore, most induction unit systems are arranged with water flowing first through the central air handling unit coils and then, in series, through the induction unit coils. Thus, chilled water may enter the central unit coil at 45° and leave at 55°, which is good for supplying the induction unit coil.

Figure 7–31 shows an induction unit system with four-pipe control. Heating and cooling coils in the primary air unit are supplied with hot and chilled water from a central boiler (or other heat source) and chiller. The primary supply air temperature is controlled at a value which varies with outdoor temperature according to some reset schedule. The chilled water leaving the cooling coil is returned to the chiller or passes into a secondary chilled water system which has its own pump and control. The supply water temperature in this secondary system is controlled at about 55° by a thermostat which operates a three-way valve to introduce water from the primary system coil or recirculate return water from the secondary system. The secondary hot water system operates in the same way.

The individual induction unit coil provides heating or cooling as required in response to the demand of a room thermostat as detailed in the discussion of the fan-coil unit. Notice that this heating or cooling may supplement or counteract the primary air action depending on the season and internal and solar loads.

Conventionally, induction units are commonly used on exterior walls and below windows; there they can furnish individual control for small zones with highly variable solar and other loads. They are not normally used for interior zones because of their small size.

A special type of induction unit is sometimes used in interior zones. This unit is installed in a return air plenum above the ceiling and utilizes the heat of the return air and light fixtures to temper the primary cold air supply for interior zone control.

7.10 UNIT VENTILATORS

A unit ventilator is similar to a fan-coil unit but is usually larger and is arranged to provide up to 100% outside air through an

Figure 7-31   Induction unit.

Figure 7-32 Unit ventilator.

integral damper system. This unit is designed primarily for heating, though some unit coils are sized for cooling also. Their primary use is in school classrooms or similarly heavily occupied spaces, where large amounts of ventilation air are required. Control systems for unit ventilators have been standardized by ASHRAE and are detailed in the *Guide and Data Book*. The principal elements of the control system are the room thermostat, discharge air low-limit thermostat, damper motor and steam or water control valve. (Electric heat is sometimes used.) When cooling is also provided, the coil control will be similar to that described for fan-coil units with the addition of outside air control. This could be as shown in Figure 7–32.

On the heating cycle the thermostat is direct-acting. With room temperature below the thermostat setting the outside air damper is closed, the return air damper is open and the normally open water valve is open. Hot water is being provided by the two-pipe system. As the room temperature increases the outside air damper modulates toward the open position. The discharge air low-limit thermostat will keep the valve open as required to maintain a minimum discharge air temperature of about 60°. When the room temperature reaches or exceeds the room thermostat setting the outside air damper will be fully open and the valve will be closed unless some heating is required to maintain the discharge air temperature low limit.

For cooling, the changeover control changes the room thermostat to reverse-acting and chilled water is supplied by the central system. Now, with the room temperature above the room thermostat setting, the outside air damper is closed, the return air damper is open and the water valve is open. As the room temperature decreases, the outside air damper modulates toward the open position and the valve modulates toward the closed position. The discharge air thermostat must be bypassed since it would try to open the valve at low discharge air temperatures, and this would of course nullify the effect of the room thermostat. Alternatively the discharge thermostat could be changed to reverse-acting, with the set point changed to about 75° as a high limit.

7.11 PACKAGED EQUIPMENT

Packaged equipment, by our definition, is HVAC equipment which is factory-assembled and ready for installation in the field with a minimum of labor and material. As such it of course includes a complete, ready-to-function, control system, usually electric but often with electronic components. Typical packaged equipment items are residential furnaces and air conditioners, rooftop units and direct gas-fired heating equipment.

Since most of these units include the controls, they will be discussed only briefly, to point out typical arrangements.

7.11.1 Self-Contained Fan-Coil Unit

A rather unusual package is the small self-contained fan-coil unit. This is basically the same as a window-type air conditioner but with the configuration of a fan-coil unit. It contains a supply fan, direct-expansion cooling coil, refrigeration compressor, air-cooled condenser with through-the-wall ducts for the air, electric heating coil, filter and controls. A return air type of thermostat is often used, so that only power and outside air connections are needed to make the unit operable. Controls are simple, electric and conventional. The thermostat is direct-acting, two-position, calling for cooling as the room temperature increases above the set point and heating when it decreases below the set point. A fairly wide differential is provided to prevent short cycling of first heating and then cooling. A manual changeover summer-winter thermostat may be used.

7.11.2 Residential Air Conditioning

A typical residential air conditioning system has a gas, stoker or oil-fired furnace within the house. This furnace has an add-on direct-expansion cooling coil installed in the discharge. Field-installed refrigerant pipes connect the coil to an air-cooled condensing unit outdoors. A two-position summer-winter thermostat with manual changeover is provided. Typically, this thermostat starts and stops the supply fan, as well as controlling the heating and cooling. A manual

Figure 7-33   Residential gas-fired furnace with DX cooling.

fan switch is sometimes furnished. On the heating cycle, the fan is started and stopped by a discharge plenum thermostat, to prevent blowing cold air. A high-limit thermostat acts to shut down the burner if the high-limit temperature setting is exceeded. Figure 7–33 shows a year-round residential air conditioning system with gas-fired furnace and air-cooled condensing unit. The thermostat has manual change-over from heating to cooling and also has an "on-automatic" fan switch. In the "cooling" position, when room temperature rises above the thermostat setting, the condensing unit and supply fan are started simultaneously, and the solenoid valve on the cooling coil is opened. The condensing unit is equipped with high- and low-pressure safety controls and thermal overload (not shown). Some units of this type and size do not use the pump down cycle. With the changeover switch in "heating" position, on a call for heat the gas burner will light (providing the pilot safety is satisfied). When the plenum temperature rises to its set point, the fan switch will start the supply fan. A high-limit thermostat (usually combined with the fan switch) will shut off the burner if necessary.

Figure 7–34 shows an oil-burning furnace. Cooling controls could be provided as shown for the gas-fired furnace. On heating, a call for heat starts the oil burner motor (or, with a gravity-type burner, opens a valve) and energizes a time-delay relay in a "stack switch." This thermostat is mounted in the combustion vent stack and must be opened by the rise in temperature of the flue gas (indicating proper combustion) before the time-delay relay times out. Thus, if the oil fails to ignite the burner will be shut off. The switch must then be manually

Figure 7-34   Oil-fired furnance.

reset. Some models of this device provide for one recycle operation to purge unburned gases and try again for ignition before finally locking out.

Figure 7–35 shows a stoker-fired furnace. The stoker-fed fire must be ignited manually, and therefore must not be allowed to go out. Because of this, the stoker control system incorporates a relay which includes a timer. Regardless of heating demand, the timer operates the stoker for a few minutes each half-hour to ensure continuation of the fire. Again, cooling could be added as shown for the gas-fired furnace.

7.11.3   Residential Heat Pumps

Particularly in the south and in apartments, small (2 to 5 ton) residential heat pumps are very popular. A typical schematic and control system is shown in Figure 7–36. Controls are, of course, part of the package. Peculiar to this equipment are the reversing valve and defrost controls.

The reversing valve is basically a four-way solenoid. It must be very carefully designed to prevent leakage between the high- and low-pressure gas streams which flow through its two sides simultaneously. The sealing problem is further complicated by the large temperature differences between the two gas streams.

Figure 7-35   Stoker-fired furnace.

Figure 7-36 Residential heat pump.

When the heat pump is operating on the reverse cycle to provide indoor heating, icing of the outdoor coil is common. Some kind of defrost cycle is therefore necessary to periodically remove the ice. Several methods are used, including:

1. A time clock provides a regular defrost cycle. This operates regardless of ice buildup.
2. A sensor notes the increase in air pressure drop through the coil due to icing and operates the defrost cycle until normal conditions are restored.
3. A refrigerant temperature or pressure sensor senses the decrease in suction pressure or temperature caused by icing.
4. A differential temperature controller senses the increasing differ-

ence between outside air and refrigerant temperatures caused by icing.

Most of these systems use air defrost, allowing the fan to run while the compressor is shut down. Units which may be required to operate in ambient temperatures below freezing must have electric heaters for defrost, or use occasional short periods of indoor cooling operation so that hot gas will defrost the outdoor coil.

7.12 OTHER PACKAGED EQUIPMENT

"Rooftop" units are packaged heating and cooling systems (or sometimes heating and ventilating only) which combine the simplicity of direct firing with the complexity of economy cycle outdoor air control, relief fans and even multizone supply. They are made in a wide variety of styles, fuels and arrangements, and with capacities up to 30 or 40 tons of cooling and several hundred thousand Btu's of heating. Controls, of course, are factory furnished and operate in many of the ways discussed in this book.

Boilers, and other direct-fired heating units, always include the controls as part of an overall package approved in its entirety by some national certification agency, such as the American Gas Association (AGA) or Underwriters Laboratories (UL). Controls on the equipment cannot be changed in any way without voiding this approval.

Common features in control systems of large gas- and oil-fired boilers are: flame-failure safety shutdown; electric ignition with checking before startup; pre-purge, to remove any unburned gases from the combustion chamber before startup; post-purge, to do the same after shutdown; and low-fire startup to insure proper operation before bringing the burner to full capacity.

Boilers of any size are often interlocked to the water pump or a water flow switch, to ensure water flow before the burner is turned on.

Makeup air units are direct gas-fired units without vents. The products of combustion are mixed with the 100% outside ventilation air supplied and are not considered hazardous because of the high dilution rate. Their use is restricted to industrial applications and commercial establishments with low personnel occupancy. Control is usually by means of a two-position room thermostat and gas valve, combined with a self-contained modulating valve with the sensor bulb in the discharge air. (See Figure 7–37.) It is often required that they be interlocked with companion exhaust systems, and in some areas they may not be used as the primary source of heating.

Figure 7-37   Makeup air unit; gas-fired.

7.13   RADIANT HEATING AND COOLING

By definition, radiant heating and cooling depend for their effectiveness on the radiant transfer of heat energy. There are many varieties of radiant heating systems, ranging from low-temperature–high-mass concrete floors to high-temperature–low-mass outdoor systems with gas-fired ceramic or electric elements. A single type of control system for all radiant systems is, therefore, not possible. For discussion purposes several distinct groupings can be recognized.

7.13.1   Panel Heating

First, and most commonly used for general heating, are floor, wall and ceiling panels. Floor panels with hot water pipes or electric cables embedded in concrete have a large mass, and therefore, a large time lag. A conventional room thermostat will not provide adequate control. A room thermostat with outdoor reset, or even an outdoor thermostat alone, has proven more satisfactory. It is essential to limit the panel surface temperature to a maximum of 85°. This can

Figure 7-38   Floor panel heating; hot water.

most easily be done by limiting the water supply temperature. Figure 7–38 shows a floor heating panel controlled by a room thermostat with outdoor reset and a high-temperature-limit thermostat in the water supply.

Wall and ceiling panels generally have much less mass than floor panels and can therefore be controlled directly by a room thermostat. While surface temperatures up to 100° in walls and 120° in ceilings are acceptable, these still require supply water temperature control. The control system would look like Figure 7–38 without the outdoor reset (although reset could still be used to improve response).

Wall and ceiling panels can also be used for cooling. Water supply temperatures must be carefully controlled to avoid surface temperatures below the room dew point. Since this type of cooling is always supplementary to an air system, the water supply can be handled in the same ways discussed for induction units.

7.13.2   Direct Radiant Heaters

Direct-fired, high-intensity infrared heaters may consume gas or electricity, and are used for general heating. They are also used for spot heating, indoors and out. Except when they are used for general heating, control should be manual, because of the inability of a conventional thermostat to sense "comfort" when heat transfer is primarily radiant.

7.14   RADIATORS AND CONVECTORS

Convectors and convecter radiators in various forms are widely used for basic heating and as supplementary heaters with air conditioning systems. This general classification includes fin-pipe and baseboard radiation, as well as the old-fashioned cast-iron column radiator and the modern fin-coil convector radiator. This equipment may use steam at low pressure or under vacuum, or hot water.

7.14.1   Low-Pressure Steam Systems

Low-pressure steam supply may be controlled by individual control valves on each radiator or by means of zone control valves which supply several radiators. In the latter case orifices are used at each radiator to ensure proper distribution of the steam, and two-position control is necessary. Steam pressure at the boiler is kept essentially constant by a pressure controller which operates the burner.

7.14.2   Hot Water Systems

Hot water may be controlled in the same manner as steam. Series flow through a group of radiators may be used if the design and selection allows for water temperature drop from the first radiator to the last.

7.14.3   Vacuum Systems

Vacuum systems are controlled by varying the steam pressure and temperature to suit the building heating requirements. In a typical system (Figure 7–39) the zone thermostat modulates the steam supply valve in response to demand. The vacuum pump is operated by a differential pressure-vacuum controller, to maintain a constant differential between supply and return mains. Outdoor reset of the zone thermostat may be provided to prevent overheating. A constant boiler steam pressure is maintained by a pressure controller. Vacuums as low as 25 in. or 26 in. of mercury are normal in mild weather. A zone may be the entire building, or individual zone supply valves and vacuum pumps may be used.

Figure 7-39   Vacuum heating system control.

Figure 7-40   Hot water supply heat exchanger.

7.15   HEAT EXCHANGERS

Heat exchangers as heat sources have been mentioned several times. Heat exchangers are required when a large central plant supplies high-temperature water or high-pressure steam, neither of which can be conveniently or safely used in the individual building HVAC equipment. As noted previously, heat exchangers may also be used to provide very-low-temperature heating water and thus protect the boiler from thermal shock and/or condensation of flue gases. Obviously, the control system varies with the function of the heat exchanger.

7.15.1   Heat Exchanger—Low-Pressure Steam

Figure 7–40 shows a simple heat exchanger control system, suitable for producing water in the 100° to 200° range using low-pressure steam or medium-temperature (up to 260°) boiler supply water as a heat source. The valve is controlled by the low-temperature supply water thermostat which may be reset by an outdoor thermostat if required. An interlock from water pump or water flow switch keeps the valve closed when water is not flowing. Modulating or two-position control may be used, with the latter usually adequate.

7.15.2   Heat Exchanger—High-Pressure Steam

When high-pressure steam is the heat source some additional considerations are necessary. The control valve must be made of materials suitable for the temperatures and pressures encountered. Relief valves should be provided on both the steam and water sides of the exchanger. And a high-limit water controller is required. (See Fig-

Figure 7-41   Heat exchanger; high-pressure steam.

ure 7–41.) Two-position control is not satisfactory due to the high temperature differentials. Modulating control must be used, and even then there is a tendency to instability and hunting unless a wide throttling range is used. Some of the problems may be minimized by using a pressure-reducing valve in the steam supply line. This makes the flow rate easier to control but the large temperature differences remain. Care must be exercised in designing and applying these control systems.

7.15.3   Heat Exchanger—High-Temperature Water

High-temperature water (HTW) poses even greater problems to the designer of heat exchanger controls. A typical HTW central plant produces water at 400° F. In order to keep the water liquid at this temperature, the system must be pressured to 350 psi or higher. These temperatures and pressures require careful selection of the control valves and piping materials.

An additional consideration is that, in order to save pumping and piping costs, the high-temperature water must have every possible Btu extracted from it at the point of use. That is, it must be cooled as much as possible. Heat exchangers are therefore selected for small final temperature differentials between the high-temperature water and the fluid being heated. "Cascading" is common, with low-temperature heat exchangers using the HTW last in a sequence which may start with low-pressure steam generators and proceed to building heating water exchangers.

Relief valves are required and control valves are commonly applied on the leaving side of the exchanger to minimize operating temperatures except for domestic hot water generators, where safety requires the control valve to be on the upstream side. Modulating control is used.

Figure 7-42   High-temperature water end-use controls.

Figure 7–42 shows a typical set of HTW exchangers for a large build-ing. The HTW temperatures shown are those that will exist at design loads and water flows. In operation they will, of course, vary some-what. Notice the "cascade" effect from the steam generator to the building heating convertor. So long as the steam generator is at full capacity, enough water will flow through valve V1 to provide full capacity to the convertor. If the convertor is at part load, some of the water will be bypassed through three-way valve V3, and the HTW re-turn temperature will be greater than 215°. If the convertor load is high and the steam generator load is low, then T2 will sense the falling building hot water supply temperature and open valve V4, providing additional HTW. Valve V4 is selected with a high pressure drop at design flow to match the pressure drop through the steam generator and valve V1.

The domestic hot water generator is not in the cascade, since its loads are so different from those of the other devices.

The manual bypass valves may be used in case of difficulty with the automatic controls.

7.16 Solar Heating and Cooling Systems

Solar energy systems come in a variety of arrangements and usages but all have three essential elements in common—collector, storage, and distribution.

7.16.1 Elementary Solar Heating System

The elementary solar heating system is shown in Figure 7-43. A water-to-water system is shown, but air-to-air and water-air systems are also used. The differential thermostat allows the collection pump to operate only when the collector temperature is at least 4 to 5° warmer than the storage. In modern water-to-water systems the collector loop will use an ethylene glycol solution to prevent night-time freezing. If this is not done, an automatic collector drain and fill system must be used, which leads to air venting problems.

The space thermostat operates the distribution pump as required to maintain space temperature, while the low limit thermostat operates the auxiliary heating system when usage exceeds storage.

Figure 7-43 Elementary solar heating system.

Figure 7-44   Self-draining collector.

7.16.2   Self-Draining Collector

The self-draining collector shown in Figure 7–44 is frequently used for heating swimming pools, where the pool itself is the storage tank. The collector is a covered tray across which the water flows in an open stream, flowing back to storage by gravity. This system can also be used for domestic water heating, using a heat exchanger as shown and eliminating the need for an ethylene glycol loop with its possible cross-contamination. The differential thermostat is the only control needed.

7.16.3   Absorption Refrigeration with Solar Heating

Figure 7–45 shows a combination heating and cooling system using absorption refrigeration. To operate efficiently the absorption refrigeration system needs water at 200° or higher, therefore a "concentrating" collector is recommended. The heating-cooling type space thermostat operates the circulating pumps and positions the three-way valves for proper operation. Notice that the three-way valves allow the stored hot water to be used directly for heating the space, or through the refrigeration system to produce chilled water for cooling the space. The absorption systems includes factory installed controls for control of chilled water temperature, condensing water interlock, and other

Figure 7-45   Absorption refrigeration with solar heating (two-pipe system).

operating functions. A two-pipe system is shown, but a four-pipe arrangement is also possible.

7.16.4   Solar Assisted Heat Pump

A very efficient way of utilizing solar energy for heating is the "solar assisted heat pump," shown in Figure 7–46. The elements of the system are the collector, heat exchanger, storage tank, package water chiller (the "heat pump") with water-cooled condenser, cooling tower for use on the cooling cycle, air handler heat exchanger coil, and the necessary pumps and valves. The table summarizes the status of the various elements on each control cycle.

Whenever the stored water is not warm enough to heat the building directly, this water is diverted to the chiller section of the heat pump where it is cooled and returned to storage. The higher temperature condensing water conveys the heat (plus the heat of compression) to the air handler. Use of the heat pump provides more storage by increasing the temperature range available and improves the efficiency of the collector by lowering the collector temperature. The system is effective with storage temperatures as low as 40°F. During the cooling season the system operates in a conventional manner, using the cooling tower for heat rejection. In the arrangement shown, the

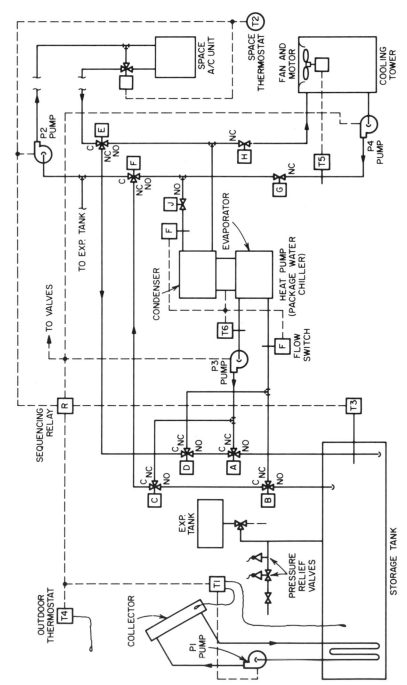

Figure 7-46   Solar assisted heat pump.

STATUS TABLE FOR FIGURE 7-46

Cycle	VALVE POSITIONS					PUMPS			Tower Fan	Heat Pump
	A–B	C–D	E–F	G–H	J	P2	P3	P4		
Solar heating	N	N	E	N	N	Note 3	OFF	OFF	OFF	OFF
Heat pump heating	E	N	N	N	N	Note 3	ON	OFF	OFF	Note 5
Cooling	N	E	E	E	N	Note 3	OFF	ON	Note 4	Note 5
Tower	N	N	N	E	E	Note 3	OFF	ON	Note 4	OFF

Notes
1. N = Normal, deenergized.
2. E = Energized.
3. Pump P2 controlled by T2.
4. Tower fan controlled by T5.
5. Heat pump controlled by T6 as a low limit controller.

cooling tower water can also be used directly for cooling. Some variations on this allow storage of cooled water by running the cooling tower at night.

7.17 Summary

In the preceding discussions the energy source has generally been ignored. Obviously, many of the functions, such as direct control of motors and solenoid valves require electrical energy. The next chapter considers the methods of describing electric control circuits and interfacing those circuits with other types of energy.

8 Electric Control Systems

8.1 INTRODUCTION

Sooner or later (usually sooner) the HVAC system designer comes to the necessity for starting and stopping electric motors for fans, compressors, boilers, pumps and accessories. And, all too often, he feels that these things are the responsibility of the electrical engineer. So, having graciously informed said electrical engineer of the location and size of these motors, the system designer goes away satisfied that he has done his part.

Unfortunately, this is not enough, as becomes evident when the equipment is installed without the necessary interlocks to the temperature control system.

Obviously, the electrical engineer must size and select the wire, conduit, starters, disconnects and switchgear necessary for supplying power and control to the motor. And this information must appear on the electrical drawings for the edification of the electrical contractor who is to install it. But these people have no way of knowing what the control system needs unless we, as system engineers, tell them explicitly and in language they can understand.

So we must learn how to express electrical equipment and wiring diagramatically using standard electrical symbols and interfacing be-

154

tween the electrical and temperature control systems with proper relays and transducers.

8.2 ELECTRIC CONTROL DIAGRAMS

Consider, then, some of the conventions and symbols used to convey electrical information. Figure 8–1(A) is a point-to-point or graphic schematic of a motor starter with momentary contact pushbuttons for manual start-stop control. The same schematic is shown in "ladder" form in Figure 8–1(B). Either of these forms is acceptable, but the ladder schematic is more frequently used and is easier to follow.

The symbol ⓘ designates the solenoil coil in the starter, which, when energized, actuates the power and auxiliary contacts. The related contacts are identified by the same number, especially when "spread out" in the ladder schematic. The symbol ⊣├ indicates a "normally open" contact. That is, when the coil is not energized, the contact is open. When the coil is energized, it closes. The symbol ⊣⫫

Figure 8-1   Motor starter with pushbuttons. (A) Pictorial diagram. (B) Ladder diagram.

indicates the opposite or "normally closed" contact. The symbol —∿∿— indicates the heater portion of the thermal overload relay in the starter. Next to each in the graphic schematic is a normally closed contact. When excessive current is drawn by the motor, for any reason, this causes the heater to heat up and open the contact. After the trouble is found and corrected, the contact may be reclosed (reset) manually. Various types of overload relays are available, depending on the size and voltage of the motor.

Notice that the auxiliary contact in the starter is wired in parallel with the "start" pushbutton. In this relation it serves to maintain the circuit after the start button is released, and is therefore called a "maintaining" or "holding" contact. When the motor stops for any reason the holding circuit opens and the motor will not restart until the "start" button is pushed.

In this and the other diagrams in this chapter the overload contacts have been shown wired on the "hot" side of the coil in accordance with recommended practice. Most starters, however, are factory wired with the overload contacts on the "ground" side of the coil. This is satisfactory if there is no ground, as may happen with some three-phase circuits. When there is a ground side to the control circuit, then the wiring should be changed to be in accordance with the diagram.

Figure 8–1 showed a typical manual system, with control power taken directly from the motor power source. Since the motor power source is frequently high-voltage (230 V or 480 V for pumps and fans, sometimes much higher for large centrifugal compressors), it is often considered desirable to provide a low-voltage source for control power. This may be 120 V or even 24 V or 48 V and may come from a separate source or from a control transformer in the starter. The advantage of the control transformer is that opening the disconnect switch interrupts all power, whereas with a separate control power source it is necessary to open two switches to interrupt all power. Of course, with complex, interlocking control systems, a separate power source is sometimes necessary.

Figure 8–2 shows a typical starter with a control transformer for supplying low-voltage control power. This is almost identical to Figure 8–1(B), the only difference being the addition of the transformer. This is necessary for proper selection of relay and starter coils.

The motor may also be controlled by a hand-off-automatic switch (Figure 8–3). Thus it may be started and stopped manually as needed, but is normally operated by a pilot-device contact in the "auto" circuit. This arrangement is used where motors are interlocked with other motors, or with temperature, pressure or flow controls.

Figure 8-2   Motor starter with low-voltage control circuit.

Figure 8-3   Motor starter with hand-off-auto switch.

Figure 8-4   Motor starter with pilot light.

"Pilot" or indicating lights are often used to indicate the on or off condition of a device. For "on" indication it is desirable to use a flow or pressure switch in the circuit to give a positive assurance of operation, since the mere fact of power being delivered to the starter does not necessarily prove that the motor is running, or, if running, is effectively moving a fluid or otherwise performing as desired.

Figure 8–4 shows the same control as Figure 8–3 but with the addition of a "running" pilot light. This figure also shows the fused disconnect switch and the motor.

8.3   ELECTRICAL CONTROL OF A CHILLER

To illustrate a relatively simple and basic interlock system consider the diagram in Figure 8–5. Here are shown a water chiller, a compressor, a circulating chilled water pump, a condensing water pump, a cooling tower fan and the necessary relays and safety controls. The power for the control circuit is obtained from a control transformer in the compressor starter, since this is the critical piece of apparatus. To place the system in operation the chilled water pump is started manually by means of a "start" pushbutton. As the water flows in the system, its temperature leaving the chiller is measured by a two-position thermostat. When the water temperature is above the thermostat setting, the thermostat contact closes, opening the solenoid valve in the refrigerant liquid line to the evaporator. The resulting rise in the suction pressure will close the low-pressure switch, starting the condensing water pump. If water is flowing in both chilled and condensing water circuits (flow switches closed) and all safety controls are closed, then the compressor motor will start.

The cooling tower fan is started and stopped by a thermostat in the condensing water supply, so that the condensing water will not get too hot or too cold.

When the chilled water thermostat is satisfied, its contacts will open, closing the solenoid valve, but the condensing pump and compressor will continue to run until the system "pumps down," that is, until the decreasing suction pressure opens the low-pressure switch. This "pump down" cycle is accomplished by the pump down relay (1CR).

Numerous safety controls are provided to protect the equipment against operating under adverse and potentially damaging conditions. The oil pressure switch contains a heater which is energized when the compressor starts. It is necessary for increasing oil pressure to open a pressure switch in the heater circuit before the heater opens a thermally delayed contact in the compressor starter circuit.

Figure 8-5 Chiller and accessories.

The compressor may also be stopped by too high a condensing pressure, too low a suction pressure, high refrigerant temperature, low chilled water temperature or inadequate flow of chilled or condensing water; or, of course, by the thermal overloads in the motor starter.

A float switch in the cooling tower sump will stop the condensing water pump if the water makeup system fails and the water level gets too low. A vibration switch will stop the cooling tower fan if the fan blades become damaged or get out of alignment.

The control system just described contains a number of conventional symbols for various types of operating and safety switches. These and many other standard symbols for electrical devices are in common use but not everyone uses the same standards. It is necessary to define exactly what your symbols mean.

8.4   ELECTRICAL CONTROL OF AN AIR HANDLING UNIT

Figure 8–6 shows a simple air handling unit electrical control sequence. Control power is obtained from a control transformer in the supply fan starter. The supply fan is started manually. Interlocks provide for operating the return fan and supplying power to the tem-

Figure 8-6   Air handling unit with return fan.

perature control system whenever the supply fan is running. The diagram shows an air solenoid for supplying compressed air to the temperature controls, but this could also be an electric relay for supplying power to electric or electronic temperature controls.

8.5 EXAMPLE: A TYPICAL SMALL AIR CONDITIONING SYSTEM

Consider now that we are designing the controls for a small commercial-type building air conditioning system, and we need to communicate with the electrical engineer.

The first step is reverse communication: We ask him what are the characteristics of the power system he is supplying—voltage, phase, cycles. He may answer that he is providing 120-208 V three-phase, 60-cycle four-wire, which gives us a choice for our motors of either 120 V single-phase or 208 V single- or three-phase. It is common practice in the industry to use the higher voltage and three-phase when available for all "large" motors. "Large" generally means ½ or ¾ hp and larger. The higher voltage and the use of three-phase power improves motor efficiency and thus lowers operating cost.

For this example our system will include an air handling unit with a 5 hp motor, a 2 hp return air fan, a ⅛ hp toilet exahust fan, a 15 hp water chiller, a 5 hp air-cooled condenser, a boiler, two 1 hp pumps and a ¼ hp air compressor for the pneumatic temperature control system.

The schematic ladder diagram would appear as in Figure 8–7. We indicate the chiller safety and operating controls by the notation "See Manufacturer's Wiring Diagram," since all package chillers are factory wired and the arrangement varies from one manufacturer to another. We might even have to correct this diagram "as-built" to accommodate the actual equipment requirements. All the other equipment layouts are similar to what we have previously discussed. The air-cooled condenser fan is interlocked with and to the chiller in much the same relationship as the condensing water pump in Figure 8–5. The boiler is assumed to be a small, low-pressure type with an atmospheric burner and a built-in factory-wired control system, for which we provide 120 V power.

It is also desirable to have a schedule of motor starters and controls, such as Figure 8–8. We fill in part of this, as shown, and give it to the electrical engineer along with the schematic. (The "NO" and "NC" contacts are the auxiliary contacts required in the starter.) He can then fill in starter sizes, make other electrical decisions, and reproduce everything on the electrical drawings. And we now have assurance that the finished job will perform electrically to the requirements of our design.

While this illustration is necessarily simple, the principles can be applied to a system of any size and complexity. Complications are limited only by the designer's ingenuity. But, as stated before, it is best to avoid complexity if possible.

If your knowledge of electric circuitry is limited, it is strongly recommended that you study of a good basic text on the subject. The more common errors among inexperienced designers are placing two or more "loads" in series (none of them work properly except in special cases) or having no load in a circuit (short circuit). The ladder schematic helps to show these errors readily. Another, not so easily seen problem is a false or "feedback" circuit, which can give full, or more often, partial control power where it is not desired. These are perpetrated even by experienced designers and arise when one circuit feeds two or more loads in parallel. If there is any question about the possibility of such a problem, use a relay or relays to isolate the extra loads.

8.6   ELECTRIC HEATERS

In Chapter 6 the control of electric heaters was dealt with in very generalized terms. Figure 8–9 shows a detailed electric control and power circuit diagram for a single-phase, single-stage heater, with manual and automatic reset high-limit controls, an electric thermostat, an airflow switch, and a relay-contactor. Figure 8–10 shows controls for a three-phase, two-stage heater, with high-limit controls, two-stage thermostat and contactors. There are many possible variations to these basic arrangements, depending on the heater size and sequencing requirements.

8.7   REDUCED-VOLTAGE STARTERS

So far in this chapter only small motors and across-the-line starters have been considered. The "inrush" or starting current on any motor is three to six times the normal running current. As motors get larger, the local electric utility requirements and good engineering practice require that we take steps to reduce this inrush current. This

Figure 8-7   Complete system electrical control.

163

Figure 8-7 (Continued)

is done by means of reduced-voltage starters. The motor size above which reduced-voltage starters are used is a function of the voltage (higher voltages mean less current for the same horsepower) and over-all size of the building electrical service. The electrical engineer can advise you of the requirements for any particular job.

Figure 8-8 Motor starter and control schedule.

Category	Mechanical Symbol	1M	2M	3M	4M	5M	6M	7M	8M	9M
	Mechanical Equipment Served	Supply fan	Return fan	Heating pump	CHW pump	Condenser fan	Compressor (refrig.)	Exhaust fan	Air compressor	Boiler
MOTOR DATA	Horsepower	5	2	1	1	5	15	1/8	1/4	—
	Voltage	480	480	480	480	480	480	120	120	120
	Phase	3	3	3	3	3	3	1	1	1
	K.W. input									0.20
	Windings									
STARTER DATA	Type						Controls by manufacturer	none	none	none
	NEMA size	1	1	1	1	1	1	1	1	—
	Number of speeds									
	Control volts	120	120	120	120	120				
	Pushbutton	S-S		S-S	S-S	—				
	Selector switch		HOA			—				
	Pilot lamp	G	G	G	G	G				
	Normally open contact	2		2	1	1				
	Normally closed contact	—	—							
REMOTE CONTROL	Control volts						120	120	120	120
	Pushbutton									
	Selector switch							on-off	on-off	on-off
	Pilot lamp							—	—	—
									Press switch	
RELAYS	PE relay number	1								
	EP relay number									
	Cont. relay									
INTERLOCK FROM			1M				5M CHW flow			3M

165

Figure 8-9   Electric heater; single-phase, single-stage.

Figure 8-10   Electric heater; three-phase, two-stage.

Reduced-voltage starters are of several types. Some require special matching motors, others work with "standard" motors. Remember that any reduction of inrush current also reduces starting torque.

8.7.1   Part-Winding Starter

Perhaps the simplest reduced-voltage starter is the part-winding starter. This must be used with a special type of motor—a part-winding motor. Figure 8–11 shows the circuit schematically, including the special motor winding. When the "start" button is depressed coil 1M1 is energized and the contacts serving part of the

Figure 8-11 Part-winding starter.

motor windings are closed. The motor starts and accelerates for a pre-set time period, as determined by the timing relay (1TR). At the end of this period the 1TR contact closes and all the windings are energized. Inrush current will be about 60% of across-the-line inrush. This starter also provides closed transition, which is discussed later in this chapter.

8.7.2 Auto-Transformer and Primary Resistor Starters

These starters may be used with any standard squirrel cage motor. Inrush may be limited to as little as 50% of maximum.

Figure 8–12 shows a primary resistor starter. When coil 1M is energized, closing contact 1M, the resistors reduce the current flow during the startup and acceleration period. When 1TR times out, coil 1A is energized, closing contact 1A. This shunts out the resistors and applies full voltage to the motor. This system may be furnished with two, three or more steps of acceleration. This is also a closed-transition circuit.

Figure 8–13 shows an auto-transformer starter. This arrangement has an excellent ratio of starting torque to power input. The "start" button

Figure 8-12   Primary resistor starter.

Figure 8-13   Auto-transformer starter.

is depressed and energizes relay 1S, which in turn energizes relay 2S. This provides current through the auto-transformer coils in the starter, starting the motor at reduced voltage and current. After an acceleration period the 2S timed opening contact opens, deenergizing 1S, and then the 2S timed closing contact closes, energizing relay R. This shunts out the transformer coils and applies full voltage to the motor windings. 2S is deenergized, restoring those contacts to their normal position with coil R held in by a maintaining contact. Note that coils 1S and R are mechanically interlocked so that only one can be energized at a time. Note also that the transformer coils are tapped for various percentages of line voltage, usually in the range from 50% to 80%. The tap selected governs the inrush and starting torque.

8.7.3 Wye-Delta Starters

Wye-Delta starters require specially wound motors, but have excellent starting characteristics, especially with very large motors. They are available with both open- and closed-transition arrangements.

Figure 8–14 shows a closed-transition arrangement. A pilot control circuit (not shown) activates control relay 1CR. Contact 1CR closes and energizes relay 1S. This in turn energizes 1M1 and the motor windings are energized in a "Wye" configuration. This allows the motor to start with low current and voltage draw. After an acceleration time period determined by the setting of 1TM, contact 1TM closes, energizing relay 1A. This first shunts a portion of the current through the resistors and then the late-opening 1A contact disconnects 1S, opening the contacts in the "Wye." The NC contact in 1S recloses energizing 1M2 and the circuit is now in running configuration with motor windings connected in a "Delta" arrangement. The pilot control circuit will usually include an anti-recycle timer which prevents restarting the motor if it is stopped within 20 min of the original start. Too-frequent starting may damage the starter.

8.7.4 Solid State Starters

A new development in reduced voltage starters is the "solid state" starter. A typical arrangement is shown in Figure 8–15. The contactors found in electro-mechanical starters have been replaced by silicon controlled rectifiers (SCR's), which provide proportional control

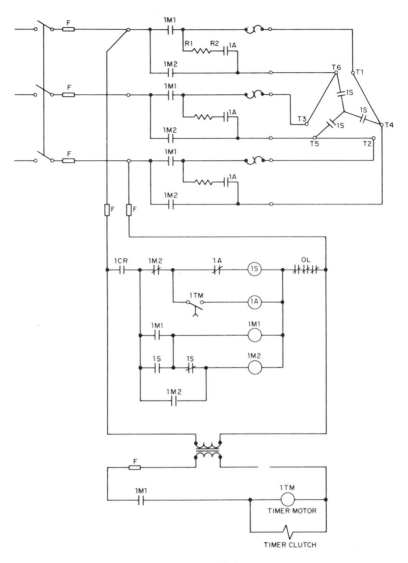

Figure 8-14   Wye-Delta starter.

of current flow to the motor. Current is sensed by current transformers (CT) (which replace the "overloads"). Current and voltage data are fed into a controller which drives the SCR's. Current during startup can be held to a desired maximum. When the SCR's are fully "on" they offer essentially no resistance to current flow and therefore act like closed contacts. Overcurrent and low-voltage protection during operation are inherent in the control logic.

Figure 8 15   Solid state starter.

8.7.5   Open and Closed Transition

All of the starter arrangements shown in the preceding paragraphs had closed-transition arrangements; that is, there was no break in the current flow during the starting sequence. The switching relay contacts were arranged so that the steps overlapped.

Open-transition starters are also available. The switching is somewhat simplified because there is a momentary break in current flow at the changeover from "start" to "run" configuration. This momentary break can cause a transient current of a very high value, sometimes called a "spike." With small motors or on large electrical distribution systems, these spikes may be unimportant. But many times a transient of this nature can disrupt an electrical distribution system, or cause a malfunction of a sensitive electronic device, such as a computer. Check this out with your electrical engineer.

8.8   MULTISPEED STARTERS

Two-speed fans are frequently encountered on air distribution systems or cooling towers. The motors commonly used in these applications are either variable-torque or two-winding motors. There are other types of multispeed motors, but they are seldom encountered in HVAC systems.

Figure 8–16 shows a variable-torque motor with simple two-speed control. Pushing the "slow" or "fast" pushbutton will start the motor at that speed. In addition to the electric interlock contacts, the two coils are usually interlocked mechanically to prevent damage to the motor.

Figure 8–17 shows the different starter arrangement used for a two-

Figure 8-16   Two-speed starter; variable-torque motor.

winding motor. The control circuit can be the same for either type of motor. The advantage of a two-separate-winding motor is that the starter circuit is a bit simpler and the motor will continue to function in an emergency on only one winding if the other is damaged. Also, two windings allow the use of any combination of speeds desired,

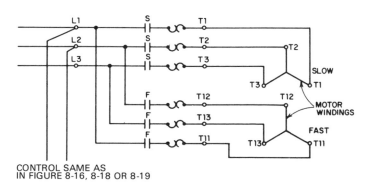

Figure 8-17   Two-speed, two-winding starter.

while a single-winding variable-torque motor must operate with high speed at some fixed ratio to that of low speed, usually 2:1.

The control circuit shown in Figure 8–16 is satisfactory for small motors. However, the same starting problems occur as previously discussed under reduced-voltage starting. A "progressive" control, as shown in Figure 8–18, has the same effect as reduced-voltage starting. When the "slow" button is pushed the motor simply starts on slow speed. When the "fast" button is pushed the motor starts first on slow speed, then after a time delay for acceleration, switches to "fast." This is an open-transition starter since it is necessary to disconnect the "slow" coil before connecting the "fast" coil.

An additional problem arises when switching from "fast" to "slow." An abrupt change will cause a dynamic braking effect, since the motor wants to run "slow" but the inertia of the fan—and this can be of great magnitude—opposes this slowdown. The resulting currents can damage the motor. It is therefore desirable to disconnect the "fast" coil and allow the system to decelerate to somewhere near "slow" speed before connecting the "slow" coil.

Figure 8–19 shows a two-speed, progressive-start, timed-deceleration control circuit. When the "fast" button is pushed the motor starts on "slow" and after a time delay switches to "fast." When this occurs, relay 2TR is energized and the 2TR contact in the "slow" circuit opens. Now, when the "slow" button is pushed, the "fast" coil

Figure 8-18   Two-speed, progressive-start.

Figure 8-19   Two-speed, progressive-start, time delay from fast to slow.

and 2TR are deenergized, but the 2TR contact provides a time delay to allow the motor to decelerate before the "slow" coil is energized. Relay SR provides a holding circuit during the delay period.

8.9   SUMMARY

The discussions in this chapter have been necessarily short. It is not the intention to present a detailed treatise on electrical engineering. It is hoped, however, that you now have a better idea of the electrical problems inherent in the design of any control system and how to cope with some of them.

9 Special Control Systems

9.1 INTRODUCTION

The control systems which have been discussed up to this point might be considered "ordinary," "run-of-the-mill" or "everyday" types. They represent the majority of the work being currently designed, and necessarily, since most air conditioning systems can be adequately controlled without frills. The basic schemes discussed can be applied in many combinations to solve the great majority of control problems.

There are, however, frequent opportunities to apply our knowledge to the solution of more exotic, complex and special-purpose problems. This chapter will discuss some of these solutions in detail, to give you an idea of the possibilities in HVAC controls.

9.2 CLOSE TEMPERATURE AND/OR HUMIDITY CONTROL

Frequently encountered are criteria for extremely close control of temperature or humidity, or both. These may and do occur in a number of unrelated end uses, ranging from machine shops to hospitals to computer rooms. Not too surprisingly, these criteria are often combined with requirements for high-efficiency filtration. Let us consider some of these applications.

9.2.1 Standards Laboratory

A standards laboratory, for either primary or secondary standard comparison, always requires extremely close control of temperature in order to maintain dimensional stability in the standard gage blocks. "Close control" here means system differentials of 0.2° to 0.05° F. Obviously, there are many special features required to make such a system controllable. Besides a high air change rate at a low differential between room and supply air temperatures, it is necessary to maintain the interior wall, floor and ceiling surfaces at or near the room temperature. This is sometimes done by using a ventilating ceiling, with a hollow wall serving as a return plenum, all heavily insulated. One excellent method involves the use of a constant cold air supply with reheat as close to the room as possible. Reheat can be by means of hot water or electric heating coils. Steam cannot be used because of its high temperature. The hot water supply temperature cannot be greater than 90°. A higher water temperature causes too great a rise in the supply air temperature, even though the control valve may be only cracked open. Electric heaters must be modulated by saturable core reactors, or solid-state controls such as silicon-controlled rectifiers.

Figure 9–1 shows a system for temperature control in a standards laboratory using hot water reheat. A special heat exchanger is required to provide 85° to 90° supply water, using the central hot water or steam as a heat source. The temperature controller must be an electronic industrial type, to achieve the necessary sensitivity. Since there will, inevitably, be some temperature variation throughout the room, the controller must be mounted as close to the gaging work station as possible, even if this means providing a pedestal or hanging it from the ceiling. The diagram indicates zone control, for one of a group of standards rooms. If 100% outside air is required, a preheat coil will be needed.

9.2.2 Clean Room

Clean rooms generally require high airflow rates to maintain the clean conditions, but temperature differential control is not usually critical. Humidity, too, is often controlled. Because the airflow

Figure 9-1 Control for standard laboratory.

rate required for cleanliness is much higher than that required for temperature control, a double air handling system may be used. Figure 9–2 shows the system and control arrangement for a horizontal laminar flow clean room. A high flow rate fan and filter system circulates the air from a plenum to the room and back again. Air temperature and ventilation rate are governed by a small air handler with heating and cooling coils and a fixed minimum outside air damper. Relief dampers are provided with static pressure control. Temperature controls are conventional, with a room or return air thermostat as the principal controller. A humidistat and humidity control equipment could readily be added.

9.2.3   Hospitals

There are a number of areas in hospitals which require close control of temperature, humidity, cleanliness or all three. These include especially: surgery and delivery rooms, nurseries, intensive care units, and laboratory and research areas. Since many present codes do not allow recirculation of air from these spaces, 100% outside air must often be used. While the laboratory area should have its own separate supply and exhaust system, the other areas can be supplied by a single air system with heat reclaim, humidity control and zone reheat. Figure 9–3 shows such a system. The outside air passes through a heat reclaim coil which is transferring heat to or from a similar coil in the exhaust air. The air is further preheated as required and passed through a sprayed cooling coil with the discharge temperature corresponding to the lowest design dew point required for any of the controlled zones. Zone thermostats control reheat coils, and zone humidistats control humidifiers in the branch ducts to those zones requiring higher relative humidities. The psychrometric charts in Figure 9–4 show the cycles for summer and winter design conditions.

9.2.4   Computer Rooms

Computer rooms have many similarities to clean rooms, but here high airflow rates are the result of high heat gains from the computer equipment. Cleanliness is not quite so important, and humidity control is required. Temperature control should be as accurate as possible (plus or minus one degree or less), since some solid-state devices are sensitive to rapid rate-of-change of temperature. Location of the

Figure 9-2 Temperature control for a clean room.

Figure 9-3 Air system with heat reclaim, 100% outside air, zone reheat.

180

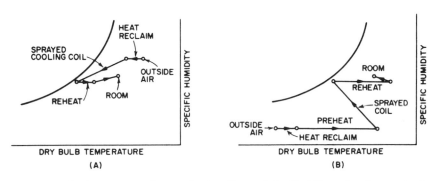

Figure 9-4   (A) Summer cooling cycle for Figure 9-3. (B) Winter heating cycle for Figure 9-3.

room thermostat and air distribution patterns become very important, since some peripheral equipments have very high heat emission rates. Some computer equipment items require direct air supply from an underfloor plenum at a controlled temperature. Other items have a complete environmental control system within them, requiring only chilled water from an outside source. Figure 9–5 shows a hypothetical system with all of these elements included. The air handling system has a sprayed coil for humidity control and separate reheat for underfloor and overhead supply ducts. A third zone supplies the office and file storage areas. Chilled water distribution is provided for "self-contained" cabinets.

9.2.5   White Room

An interesting control application occurs in a machine shop "white room." Such a room, while not a "clean room," has a high degree of cleanliness. The machining operations carried on require extremely close control of temperature, and a fair degree of humidity control. Since the room must be large to accommodate machining operations, it is impossible to provide a single location for the thermostat which will adequately control the environment at all the machines. But only one or two machines are critical at any particular time. An industrial quality electronic controller may be mounted on a rolling pedestal with trailing cable connections. The controller may then be relocated anywhere in the room, as desired.

Figure 9-5   Computer environmental system.

9.3   CONTROLLED ENVIRONMENT ROOMS FOR TESTING

Environmental test rooms are used in industry for testing products. Prefabricated chambers complete with controls may be purchased from several manufacturers.

Environmental test rooms for physiological testing of humans are usually custom designed as part of a laboratory complex. The air conditioning system is required to provide a wide variety of environments, for example, from 32° to 120° F and simultaneously from 20% to 80% relative humidity, with close control of differentials at any set point. Obviously, a complex system is required.

Figure 9–6 shows an air conditioning and control system which will satisfy the requirements of a physiological environmental test room. The small amount of outside air required for ventilation is mixed with return air and then flows through a cooling coil. This is a direct-expansion coil, with its own air-cooled condensing unit, and with widely spaced fins to minimize the effect of ice buildup at low temperatures. The coil is provided with an automatically adjustable back-pressure suction valve which is operated by the room thermostat to control the coil suction pressure and thus the minimum temperature of the air leaving the coil.

The air then passes through the supply fan and to a chemical dehumidifier which is provided with a bypass and a regeneration air duct and fan. The dehumidifier includes circular "trays" of a chemical absorbent which rotate through the supply air and regenerant air in series. The supply air is dried, and the heated regenerant air removes the moisture and carries it away. An electric heater and duct thermostat are provided for controlling the temperature of the regenerant air.

The humidistat controls the proportions of air flowing through the dehumidifier and bypass since direct control of the dehumidifier is not easy to accomplish. The supply air leaving the dehumidifier is hot, due to the heat given up by the hot absorbent as well as the heat due to the chemical reactions. It must, therefore, be recooled, and a second cooling coil and condensing unit are provided for this purpose. An electric reheat coil provides final temperature control when high temperatures are needed. A steam humidifier provides high-level humidification.

The basic thermostat and humidistat are recording-type industrial electronic instruments with remote sensors, so that remote monitoring and adjustment of room conditions are available.

Figure 9-6 Environmental test room:

184

9.4   SOME EXAMPLES FROM PRACTICE

The next paragraphs will describe some rather complex HVAC and control systems taken from the archives of Bridgers and Paxton, Consulting Engineers, Inc. and used with their permission. The layouts have been somewhat simplified but all of the essential elements are retained. These examples illustrate some of the problems encountered in practice and their practical solution. You will note that, complex as these systems are, the basic elements are the same as those that have been discussed throughout the book. These elements are simply combined in new and ingenious ways to solve specific problems.

9.4.1   Office Building with Water Source Heat Pump

The Simms Building (Figure 9–7) is a 12-story office building, with the long axis east-west and large glass areas on the north and south faces. The thermal system is a heat pump with water from two wells as both source and sink. The south-facing exterior zones have such high solar loads that cooling is required even on cold days (as low as 19° F outside). The basic air conditioning is provided by multizone units, with supplemental heating and/or cooling provided by wall and ceiling panels at the north and south exteriors. Because of the low-temperature heating water available from the chiller condensers (115° F maximum), heating coils are much larger than normal and hot water flow, as well as chilled water flow, to the multizone unit coils is uncontrolled. Panel water supply temperatures are carefully controlled on a schedule set by an outdoor master thermostat. The well pumps are operated manually so that the shallow well is used for heating supply and the deep well for cooling supply. Valves in return water to wells are manually positioned so that return water goes to the well which is not pumping.

Pump P1 takes the well water from the settling basin and pumps it to the evaporator and/or the condenser, depending on the position of valves V1 and V2. V2 is controlled by a high-limit thermostat T2 in the HWS leaving the condenser, and opens and makes up cooler well water to the condenser to avoid excessively high condensing temperatures. V1 is controlled by a low-limit thermostat T1 in the chilled water leaving the evaporator and makes up "warmer" well water to avoid

Figure 9-7   The Simms Building.

Figure 9-7 (Continued)

freezeup. Submaster cooling and heating thermostats T3 and T4 control the chiller compressor through relay R10, to maintain design hot and chilled water temperatures. The set points of these two thermostats are reset by outdoor master thermostats T5 and T6. Thermostat T6 also controls a complex series of logic relays to provide control air to the heating or cooling controls as required by the season. Relays R4 and R5 operate to limit well water makeup to that required by thermostat T2 when outdoor temperatures are between 50° and 65°. (V1 is closed by means of R3 and R1.)

Pumps P2 and P3 are the primary pumps for the chilled water and hot water circulating systems, respectively. Hot and chilled water circulate through the multizone unit coils without control at a preset design flow rate.

Pumps P4 and P5 are secondary pumps for the panel water circulating systems. Valves V3, V4, V5 and V6 are positioned by outdoor thermostat T8 (through relay R11) so that chilled water is supplied to the panel system when outdoor temperature is above 60° and hot water is supplied when the outdoor temperature is below 60°. Three-way modulating valves V11 and V12 are controlled by submaster thermostats in the panel supply water, with supply temperature set points reset by outdoor thermostats. The south panel system outdoor thermostat is "solar compensated," to take account of the solar load.

Any well water which is added to the system must be returned to the other well, and this is accomplished by means of valves V7 and V8 for chilled water and V9 and V10 for hot water. These valves are opened in sequence by pressure controllers.

The multizone air handling units have conventional mixing damper controls with zone thermostats. Minimum outside air dampers open when the supply fans run. Maximum outside air and relief dampers open only between 60° and 75° outdoor temperature. The economy cycle is not used because at lower temperatures the system needs to utilize internally generated heat to minimize the use of well water. At ideal conditions the entire system is self-balancing, that is, no well water is needed and heating and cooling loads are in balance.

The diagram has been greatly simplified (as compared to the original) to make it more easily understood. There are actually two pumps each at P1, P2 and P3. There are two chillers in the heat pump system. The compressor control system, not shown, consists of a set of sequenced switches to start and stop compressors and load or unload cylinders in response to demand. Service and bypass valves are provided extensively throughout the system.

9.4.2   Life Science Building

A college laboratory and office building for the study of "life sciences," that is, biology, zoology, etc., presents a completely different problem. Here there are high volumes of exhaust, some from fume hoods with high-pressure drop in the hood, and the rest as general exhaust with low-pressure drop through a grille. Static pressure controls must be provided to keep fumes and odors from traveling from the laboratories to adjoining corridors and offices. Each laboratory must have its own temperature and, in some cases, humidity control.

Figure 9–8 shows a solution to these problems. A single, large, variable-volume-with-reheat system serves the entire building. A single exhaust system serves all the exhaust requirements. A booster exhaust fan is provided at each fume hood to overcome the hood pressure loss. A runaround heat reclaim system is provided. This serves to partially preheat the 100% outside air supply. A preheat coil with circulating pump provides the rest of the preheating to 53°. In summer the runaround coil system provides some precooling and the cooling coil provides additional cooling to 53°. Total air supply varies according to demand and is controlled by a static pressure controller which senses "end-of-main" pressure and modulates inlet vane dampers on the supply fans to maintain this at +0.25 in. of water with respect to outdoors. Modulation of the variable-volume supply dampers will cause a change in the "end-of-main" pressure. Each room supply is provided with a variable-volume damper and reheat coil. These are controlled by the room thermostat as described in paragraph 7.5.1, except that when the fume hood exhaust fan is started (manually) a solenoid relay causes the supply damper for that room to go full open. Another solenoid relay opens the damper in the hood exhaust duct and closes a damper in the room "general exhaust" duct. A pressure sensor in the exhaust branch main to the hood will stop the fume hood fan if the branch main pressure becomes greater than −0.10 in. of water. The exhaust system is maintained at an end-of-main negative pressure of −0.25 in. of water with respect to outdoors by a static pressure controller which modulates inlet vane dampers on the main exhaust fans. This combination of pressure sensors maintains a pressure gradient from supply to exhaust which effectively prevents exfiltration from the laboratory areas.

Some rooms require humidity control, and this is provided by steam humidifiers in the room supply ducts, with control by room humidistats.

Figure 9-8   Life science building.

SUPPLY AIR SYSTEM

LABORATORY TEMPERATURE
CONTROL
(ROOMS WITH FUME HOODS)

Figure 9-8 (Continued)

Environmental rooms in which temperatures may be varied from 60° to 90° F are provided with high-capacity reheat coils and unit coolers.

9.4.3   Office Building with Internal Source Heat Pump

An office building for an electric utility company had as a basic criterion that electricity should be the only energy source. Studies indicated that an internal heat source heat pump would be the most economical solution. Figures 9–9 and 9–10 show the highlights of the resulting system.

The air system, Figure 9–9, consists of two main systems. One of these provides cooling only and serves the interior zones, which, of course, require no heating. This system includes return-relief fans and economy cycle outside air control, with the added feature of nighttime shutoff of outside air. The cooling coil valve is controlled by a discharge thermostat to provide 50° F supply air. Individual rooms or zones are provided with variable-volume dampers. A static pressure controller in the supply main senses the pressure changes as the variable-volume dampers modulate and adjusts the supply air volume accordingly by means of inlet vane dampers on the supply fan.

The other air system is a dual-duct, high-velocity type, again with return-relief fans and special outside air control. Notice that the low-limit "mixed-air" thermostat is in the cold plenum and senses the temperature leaving the cooling coils, rather than the true mixed-air temperature. Since the cooling coils are uncontrolled, this has the effect of limiting the outside air quantity in mild weather, allowing the use of internal heat for heating exterior zones, rather than dissipating it in relief air. Night shutoff of outside air is also provided. The supply and relief fans are provided with dampers which close when the fan is off, thus preventing short-circuiting of air when not all the fans are running.

Perimeter walls are provided with a continuous series of induction units and these are served by zone air systems supplied through dual-duct mixing units. The induction unit coils are controlled by means of the zone thermostat and four-pipe control valves. The mixing unit pressure controller is set to provide a relatively high outlet pressure, to insure proper operation of the induction units.

Mixing units with low outlet pressures are provided for the overhead air distribution systems which supply most of the air to the exterior and public zones.

Figure 9-9 Office building with internal source heat pump—air systems.

193

Figure 9-10   Office building with internal heat pump—water system.

Figure 9-10 (Continued)

Figure 9–10 shows the system for providing hot and chilled water. Three package water chillers are arranged in "cascade" fashion, that is, the chilled water and condensing water flow through them in series and with counter flow. Thus, each chiller operates on about the same temperature difference between condenser and evaporator. The cascade system makes possible greater overall temperature changes in the hot and chilled water, allowing smaller water quantities to be used. At light loads one or two of the chiller units may be shut down manually, but normally all units are running and capacity is controlled by relay R12, which selects the greater of the two pressures fed to it by the reverse-acting hot water and direct-acting chilled water temperature controllers, T3 and T4. Output pressure from these controllers flows through relays R10 or R11, which are positioned by the manual selector switch. Below 45° outside, "heating" is selected and the hot water controller controls. Above 60° outside, "cooling" is selected and the chilled water controller controls. Between these two temperatures, "auto" is selected, and both controllers are effective. The one with the greater demand controls the chiller capacity. Also, in the "auto" position, changeover may be automatic as described below.

In addition, these two controllers operate through differential pressure control P1 as follows: If the "cooling" branch pressure is greater than "heating," the output of P1 is equal to the difference in pressures. If the "heating" branch pressure is the greater, the output is 0 psi. A 7 psi output pressure changes relay R4 to provide a 15 psi output which starts the cooling tower pump (through PE5) and positions relay R8 to allow the hot water high-limit controller T2 to operate (provided the manual switch is on "cooling" and R9 is in the normal position). Also, valves V1 are positioned to circulate the hot water through the heat exchangers and thus dissipate excess heat to the cooling tower. Thermostat T2 acts to start and stop the cooling tower fans and position the tower water bypass valve, V2, in accordance with demand

When the output of P1 is 0 to 3 psi, the cooling tower and pump are inoperative and valves V1 are positioned to recirculate the hot water and bypass the heat exchangers. T2 is also inoperative since it is assumed the heating load is sufficient to keep the hot water cool enough to prevent excessive condensing temperatures.

The hot water temperature controller T3 is reset by an outdoor master thermostat, which also operates through relays R1 through R6 to provide control air supply to T2, T3 and T4, thus: When the outdoor temperature is below 45° the branch pressure from T1 is low, all relays are in the normal position and heating control only is provided.

Control air is supplied through R3 and R5 to the hot water submaster controller T3, which is reset to some high control point. As the outdoor temperature (and T1 branch pressure) increase, relay R1 switches, switching R5 and R6 which provide control air for both heating and cooling. On a further increase in outdoor temperature, R3 and R4 are switched, but outputs of these relays are blocked by R5 and R6. At about 65° relay R2 is switched, returning R5 and R6 to normal and providing control air for cooling only through R4 and R6. As the outdoor temperature falls, a reverse sequence takes place.

The induction unit chilled water pump is manually started. Thermostat T9 in the induction chilled water supply positions three-way valve V3 to prevent too low a temperature of the supply water. Differential pressure control P2 modulates valve V4 to maintain a maximum differential pressure from supply to return. This is effectively a bypass valve, to maintain flow if the induction unit control valves are closed. A similar control is provided in the induction unit hot water system.

The induction unit hot water pumps are started manually. No control is normally provided in the hot water supply, but a low-limit thermostat T10 may bring on an electric boiler and circulating pump if the temperature falls too low at night.

A low-limit thermostat T8 in the chilled water supply operates through relay R13 to turn on portions of the building lights, thus creating a false load and maintaining a minimum chilled water temperature.

9.5   SUMMARY

It is obvious that the systems described above required a great deal of thought and experience to design. In any case, however, the overall system is built up from relatively simple subsystems, and it is their interplay which creates the complexity. Careful study of these designs will be rewarding and educational. Some of these arrangements may appear to violate the rule of simplicity, and perhaps you can find a simpler way to accomplish the same program.

10 Supervisory Control Systems

10.1 INTRODUCTION

The increasing size of modern buildings and building complexes, and the difficulty of obtaining competent operating personnel have led to an increasing use of central supervisory control systems. These systems allow one person, at a central location, to monitor and control the operation of up to several thousand elements of the heating, ventilating and air conditioning systems in the building or complex. The addition of alarm circuits and audio or audio-visual communications improves security as well as simplifying and improving the degree of control which can be maintained. The additional first cost of such a system can usually be amortized in a few years by the savings in personnel costs and more efficient operation. Where numerous tenants are involved, the increased tenant satisfaction may be sufficient justification for the added cost.

10.2 HARD-WIRED SYSTEMS

"Hard-wired" central systems were the first used. As the name implies, this is simply the extension of the conventional individual control wires to a central point. Each element in the control system, whether start-stop control of a motor, indication of a temperature, reset of a control point or malfunction indication requires one or more wires running all the way to the control panel, plus a separate device on the panel. Obviously, a building need not be very large before this becomes too cumbersome to be useful. Many such early systems were built and are still in use.

The transmission of signals from distant air handling units and other equipment to the central control point poses some problems. Two-position or digital signals are simple. These are used for starting and stopping motors and indicating on-off or out-of-control conditions. Analog signals for measuring flow, pressure, temperature and the like are more difficult. These signals may be transmitted as low-voltage direct-current signals, or as air pressure signals which can be translated and read at the receiving end by means of a transducer. In either case, line losses due to distance become important, and careful calibration and compensation is necessary.

10.3   MULTIPLEXING SYSTEMS

A complex involving a great many similar control systems such as air handling units can be more efficiently and easily handled by the use of multiplexing relays. A multiplexing system provides one set of channels to cover the required control functions which are common to all units. A relay at each individual air handling unit allows that unit to be selected for monitor and control through the common channels. When the routine for one unit is completed, another unit can be selected, with the same common channels used. The relays used are called multiplexing relays. The control and monitor wiring is greatly simplified. After basic wiring is provided, we need only one additional wire for each additional station. Figure 10–1 shows the elements of this system in simplified form.

When the selector switch is positioned for a particular system, coil no. 1 in the multiplexing relay is energized, closing contacts no. 1 and putting the system "on-line" to the control elements at the central panel. Then the motor may be stopped or started from the central panel, with pilot light indication provided. In the wiring shown here, local start-stop pushbuttons and pilot light are also provided.

Analog functions, such as temperature indication and reset, can be added by using a different type of relay.

For additional information, a projection screen may be provided,

Figure 10-1 Central control using multiplexing. (Courtesy Honeywell Inc.)

Later systems of this type use coding signals to energize the relay. This requires only a few wires to address a large number of relays.
with an automatic projector and slides to give the operator a visual picture of the system with which he is dealing.

Alarms for system malfunction must be hard-wired, or they can be superimposed on the basic system wiring using various proprietary devices based on solid-state generated high-frequency signals.

Where the equipment rooms are remote from the central control panel, as is usually the case in a large complex, the control system may include an audio or audio-visual communication system. With this available, the actions of a maintenance man doing trouble-shooting on the equipment may be guided by the supervisor at the central location. Closed-circuit TV can also improve security by providing visual observation of unoccupied equipment rooms.

10.4 COMPUTERIZED SYSTEMS

A natural extension of the central supervisory control system is to replace or supplement the operator with a computer. *Computer-type* and *computerized* supervisory systems are essentially similar, differing mainly in speed and flexibility. The term "computer" is usually understood to mean a general-purpose programmable computer. As used in the control industry, it is also a "mini" computer, designed especially for process control (as contrasted with a data-processing computer which is usually large and faster). (Figure 10–2)

The typical mini computer uses a 16-bit word and has a "core" storage (in the computer itself) of 32,000 words or more. As noted below, additional storage is available in peripheral equipment such as disc and tape machines. Speeds are relatively slow with cycle times of 1 to 3 microseconds. Since this is several times the speed of the communication system through which the computer operates, faster cycle times are unnecessary. These computers are easily programmed, using Fortran or some similar language, thus they are very flexible and can be adapted to any installation.

"Computer-type" systems are designed around a device that is sometimes called a "processor." This is also a computer, but one designed for the specific purpose of operating a supervisory control system. It is usually programmable only by special methods and lacks some of the flexibility of the general-purpose computer. However, it is somewhat less expensive and because of its simpler design may be easier to use in some circumstances.

Figure 10-2 Typical mini-computer. (*Courtesy Powers Regulator Co.*)

"Micro" computers are sometimes used (Figure 10–3). These are, as the name implies, smaller than mini computers but have surprizing capabilities. An entire micro computer, lacking only a power supply and a console, can be contained on a small printed circuit board. Re-programming capability is limited, since memories are of the "read-only" (ROM) type.

10.4.1 Computer System Components

In addition to the processor the basic elements of the computer-type system are the *operator's console, peripherals* such as printers, CRT's and mass storage, *graphic displays, the communication system* and, finally, the *sensors and controllers* (Figure 10–4).

10.4.1.1 Interface Device The interface device (ID) (Figures 10–5 and 10–6) between the communication system and the sensors/

Figure 10-3 Micro-computer. This single circuit board contains all the computer functions except power supply. *(Courtesy Honeywell, Inc.)*

controllers is given various names by different manufacturers, but its functions are always the same. The basic functions are: address recognition, buffering, timing, analog to digital (AD) conversion, (and sometimes digital to analog conversion), incoming signal conditioning, multiplexing, and outputting control signals.

The sensor and controller points in one location in a building (i.e., a fan room) are connected to the interface device serving that area. The addresses of these points can be recognized by the ID and in passing information on to the communication system it attaches the address data. Commands for control points are routed through the proper ID by means of these addresses.

All data passing through the ID must be *buffered*. A buffer is a storage area where the data are retained until transmitted or otherwise operated on. Electronically a buffer is a number of bits arranged in a specific manner. This is similar to computer storage methods.

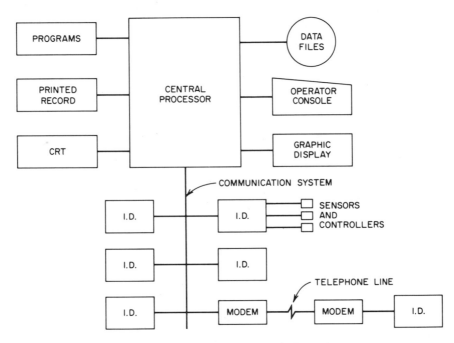

Figure 10-4   Computer control schematic.

Many ID's include *timing* functions. That is, the ID has within it a clock which provides for performing certain specified operations on a regular schedule. A typical timing operation is that of scanning all points connected to the ID and storing, transmitting, or comparing the data.

Analog to digital (AD) conversion is essential with digital communication systems. The convertor takes the analog data from, for example, a temperature sensor, and changes it into a digital array, which can then be transmitted as a group of bits. For process control systems an 8-bit group gives a sufficient degree of accuracy (one part in 256). Figure 10–7 shows a typical 8-bit group. The computer handles this as a group in which each bit may be on or off (1 or 0) but in the group each bit has a different significance, increasing from right to left by powers of 2. In the figure, the value of the group is $16 + 8 + 2 + 1 = 27$. If all bits are 1 the value of the group is 255.

Digital to analog conversion works in a similar manner, and the analog output may be used to reset a control set point, to position a damper or valve, or some similar analog control function.

Signal conditioning is the process of converting a signal from an external source to a type and level which is suitable for the ID and communication system. For example, a pneumatic sensor provides a signal of varying air pressure. The ID usually needs a low voltage electrical signal. A signal conditioner or transducer will be provided to convert the pneumatic signal to an equivalent electrical signal. More often, the transducer simply changes an incoming high voltage signal to the lower voltage, or converts a varying amperage signal to an equivalent voltage. A simple two-position (contact closure) signal needs no con-

Figure 10-5 Typical "Interface Device." This is a large unit with a power supply in the top section and several function cards in two card racks below. The gages indicate output pressure from variable pneumatic controllers. *(Courtesy F & M Systems.)*

Figure 10-6a Function card. A detail of one card from the rack of Figure 10-5. *(Courtesy F & M Systems.)*

Figure 10-6b Function card. *(Courtesy F & M Systems.)*

Figure 10-6c   Function card. Analog to digital conversion. *(Courtesy F & M Systems.)*

ditioning. By use of the proper transducers, almost any type of sensor may be connected to the ID. It is preferable, though, to use the control system manufacturer's standard sensors when possible, to avoid the expense of special transducers.

Multiplexing is similar to the process described previously, but is done within each ID as well as in the main communication system. By means of an internal multiplexer, the ID addresses each of its connected points in turn, sending or receiving signals and processing the signals in a prescribed sequence. The central processor, through its operating system, controls the flow of traffic on the communication system and determines when each ID will be allowed to send and/or receive. Nor-

Figure 10-7   8-bit group.

mally all points are scanned in a routine sequence, but *interrupt* capability is provided to deal with alarms. In effect, the ID says "Help, I have an alarm," and in response the processor says, "O.K., tell me your trouble." The alarm is communicated and processed, and the routine scan is resumed. Alternatively, all alarm points are scanned at intervals of 2 or 3 seconds so that there is no delay in sensing alarms.

Signals are received by the ID from its various sensors and alarms, and control signals are output. These may be analog, as described above, or two-position—closing or opening a relay. Two-position controls are used to start or stop motors, to open or close valves or dampers, or to drive a varying position device such as a potentiometer for some set time (thus, in effect, providing an analog output).

10.4.1.2 **Operator's Console**   The operator's console is the essential link between man and system. By use of the console the operator may obtain information, issue commands, and override the automatic programs which normally control the system.

The console is a keyboard: either a typewriter keyboard with special keys added, or special keys only. These special keys are designed to simplify the operator's task and have names such as "next point" (call up the next point in the address table), "last point," "set point limits," "alarm summary" (type out a list of all alarms in the last period), "all points" (log the status and/or value of all points in the system), "start-stop," and so on, to the practical limit of the console designer's imagination and the available funds.

Figure 10-8   Operator's console. *(Courtesy Powers Regulator Co.)*

Figure 10-9 Operators console. This is two consoles, one for HVAC and one for fire and security, with a typewriter. These have CRT displays. *(Courtesy F & M Systems.)*

The console usually includes a display showing the name and value, with engineering units of the particular point being addressed. Since only one point may be displayed at a time, auxiliary graphic displays are usually used to set the point in context.

10.4.1.3 Graphic Displays are used to assist the operator in visualizing the systems and points within systems. The slide projector displays developed originally for electromechanical supervisory systems are used extensively in computerized systems. Here, however, it has been possible to also display a list of points together with their changing values, either directly on the projection screen or adjacent to it. The more sophisticated systems use a CRT (cathode ray tube) display. This is simply a small TV screen which can be made to display a list of points and their values (Figure 10–10). The CRT can also be used as the operator's console, in which case it has a keyboard and displays the operator's commands and system responses. Finally, a black and white or full color CRT can display a picture of a system similar to that shown on a slide projector (Figure 10–11).

All graphic displays are controlled to automatically display the sys-

Figure 10-10   CRT display. *(Courtesy Powers Regulator Co.)*

Figure 10-11   CRT graphic display. *(Courtesy F & M Systems.)*

tem containing the point being addressed, or, when an alarm comes in, the display shows its location, often with a flashing red light. The advantages to the operator are obvious.

10.4.1.4 Logging Most modern supervisory systems utilize "logging" typewriters. These automatic machines keep a permanent record of all system activities, including alarms, operator commands, point status, and other operating data. These records are useful as history, for analysis, and sometimes for legal purposes. The typical entry includes data, time, point address, and what happened—recording of value, status or alarm or issuance of or response to a command. More sophisticated systems can be programmed to store and analyze data and print out summaries.

Another use of the typewriter is in printing out maintenance instructions. These instructions are stored by the computer, to be issued as required by the operating time of the various pumps, fans, chillers, etc. in the HVAC system. After the maintenance is performed the operator informs the computer and this becomes part of the system history.

10.4.1.5 Communication System The communication system is the link between the computer and the interface device. (In a sense it also includes the wiring between the sensors and controllers and the ID, as well as the ID itself). This link in computer-type systems consists of a coaxial cable, or one or two pairs of wires ("twisted pair" is a term heard frequently) (Figure 10–12). In some systems the link forms a complete loop from the computer through all the remote ID's and back to the computer. In other systems a direct "out and back" connection is used. In either case service to the individual points and ID's is provided by means of time-sharing. The messages to and from follow one another in a continuous stream, and each includes the address to which it is directed (or from which it emanates) and the data or instructions. It is the responsibility of the computer to control and sort out all the traffic, giving priority to alarms or critical instructions and maintaining a routine flow when everything is normal.

Virtually all modern communication systems use digital transmission rather than analog. The disadvantages of analog transmission are: 1. Attenuation of the signal with distance, so that amplifiers are necessary for distances of over 1000 feet or so. 2. Susceptibility to "noise," that is, stray electromagnetic currents which are broadcast by many types of equipment and picked up by the antenna effect of communication wire. Noise can be filtered to some extent, and accuracy of the message is checked by transmitting it two or three times. There is, however, general agreement that digital transmission is more reliable.

Figure 10-12 Communication cable. Left to right: 1 coaxial cable, 2 twisted pair with ground wire, 3 twisted pair with shield. *(Courtesy Powers Regulator Co.)*

A digital signal is "on" or "off." When message blocks of 16, 24, or 32 bits are used, the value of the signal relates the message. In Figure 10–13 is shown a typical message format. Since the value of each bit is always 1 or 0, attenuation is not important and noise problems are minimized.

Some systems use frequency, a digital system which is not in unique "bits" but in which the distance between signal "ones" is the message value.

Modems. A "modem" is a device which allows the communication system to interface with an ordinary "voice grade" telephone circuit, for transmission over long distances. Another modem at the other end of the telephone link connects with the communication systems and interface devices in the far-off building. The telephone line is permanently connected on a lease basis. By this method one supervisory panel can be used to control 2 or 100 or more buildings scattered over a wide geographical area.

ADDRESS	MESSAGE	PARITY

Figure 10-13 Typical message format. The "parity" bits are used for checking accuracy of transmission.

Voice communication is often used with any type of supervisory system. The intercom system requires its own pair of wires but offers many advantages. It allows the maintenance man in a remote machine room to discuss his problems with the central plant operator. It also allows the operator to "listen in" on the remote room to hear motors actually start or stop when commanded, or to listen for inappropriate noises which indicate equipment malfunction. The control plant operator can select any or all of the remote stations, while from the remote station it is necessary to use a "push to talk" switch, unless the speaker/microphone is "voice-actuated."

10.4.1.6 Sensors and Controllers The sensors and controllers used for supervisory control are usually separate from the devices used for the local closed loop control. They also tend to be electronic because it is easier to interface with the supervisory system. This is not, however, essential, given the proper signal conditioners and transducers. The distance between analog sensors and the interface device should be as short as possible due to the attenuation of analog signals in transmission and the possibility of picking up "noise" (stray electromagnetic signals). The ideal analog sensor will have the AD conversion done at the sensor to avoid the problem. Attempts are being made to develop sensors of this type.

Some people have advocated using the computer to close the local loop. Refer back to Figure 1–1 and put the "computer" in the position of the "controller" to understand the proposal. While this type of arrangement would function satisfactorily, the loss of the computer or any part of the communication system would cause the entire control system to fail. When local loop controls do the actual controlling under the supervision of the computer system, then the loss of the supervisory system is not so critical. Most manufacturers favor this latter approach.

10.4.2 Security and Fire Reporting

Security and fire reporting functions are fundamentally the same monitor-control-alarm functions used in HVAC supervision. The systems which are used for HVAC control can therefore easily be adapted to these additional functions. In some localities code authorities require the use of "proprietary" and UL labeled systems for fire reporting. The basic HVAC system with a few special features will meet this need.

Fire reporting systems, particularly in high-rise buildings, require a special control panel for fire department use. The computer-type system is well fitted for this arrangement, with the special panel protected against unauthorized use by a key lock or special code. Many of the HVAC control manufacturers have a version of their HVAC supervisory control adapted especially for fire and/or security use.

Security systems get into many areas, such as door locks activated by magnetic cards, intrusion alarms, and proximity sensors. Closed-circuit TV can also be used, though this is not a part of the basic supervisory system.

10.4.3   Software

The function of the computer (or processor) is to supervise the operation of the system. It provides the necessary programs, timing, addressing, response to data, analysis and interpretation of operator input, and control of the communication system. These functions are performed by means of *software*, that is, the *programs* which the computer follows. A program is a step-by-step set of instructions which direct the computer in each of its actions. These instructions must be carefully written in a special language (which is different for each make of computer). For example, pressing a function key on the operator's console will trigger a lengthy instruction set, each step of which is performed in sequence with the desired end result obtained only if each step is correctly written and performed. The standard programs available from each control manufacturer have been carefully written and thoroughly tested for errors. This testing process is called "debugging" and can sometimes be very frustrating to the programmers and system analysts who are responsible for software development. "Customized" programs, written especially for a particular project can become very expensive. "Standard" software which has already been written and debugged is comparatively much less costly.

Programs fall into several classifications. *Operating* programs perform the various functions for which the system is designed (control, monitoring, etc.). The *Data base* is a list of all points in the system, together with the address and data pertinent to each point. *Diagnostic* programs continually survey the computer system elements, diagnosing and reporting hardware problems. *Executive routines* are the basic programs which tell the computer how to perform each of the operations necessary for proper functioning.

The processor is provided with several *channels* through which it interfaces with the peripheral equipment and the system communication link. More channels allow greater speed and flexibility in the system, but are also more costly. Six or seven is an average number.

Most control systems manufacturers now offer "modular software." These programs are written to cover specific functions such as timed start-stop of equipment, data acquisition and analysis, and "optimization" of various kinds. The programs are designed to operate on any system by accepting the data base for that system. This kind of software is comparatively inexpensive and allows the operator to take advantage of the computer's flexibility in ways which are beyond the capability of the simpler supervisory systems.

When the number of points in the system reaches 400, 500 or more, or when optimization programs are used, the memory in the computer becomes too small to store all the necessary data and programs. Then a "mass storage" peripheral is added, usually in the form of a high-speed "disc" which can store several million bits of information. The disc consists of several (4 is typical) plates, about the size and appearance of large phonograph records, rotating at high speed on a spindle. Between each pair of plates and above and below the top and bottom plate are fixed rectangular "heads," spaced so they do not touch the plates. The data are read onto or off the plates by means of the heads, through an induction principle, and the high rotational speed allows very fast access to data, on the order of a few milliseconds (as contrasted

Figure 10-14   Cassette tape for program input. *(Courtesy Powers Regulator Co.)*

with microsecond access to data in the internal memory of the computer). Access to the correct data is provided by a complex addressing function.

Tape machines are also used for mass storage. These are used extensively with scientific processing computers, but not so often with process control computers. The tape is similar to, but much larger than, the tapes used on tape players for audio recording. Tapes are very useful for permanent storage of data; when one tape is full it can be removed from the machine and stored. For most process control systems a small cassette tape is used to input programs. The programs are compiled on a large computer by means of punched cards, then stored on the cassette tape, for play back into the process control computer (Figure 10–14). Major program changes are handled in this way, though most systems have a limited programming capability from the operator's console. The term "keyboard programmable" usually refers to this ability to adjust limits, add or delete items in the data base, and similar functions which do not change the basic programs.

10.4.4   Other Terms

Some other terms which may be encountered are:

1. *Nonvolatile Memory.* This refers to a computer memory so designed that in case of power failure the current data in memory are not lost. When power is restored the computer can proceed from where it left off. In some systems a battery is provided to allow enough time for an orderly shut-down with memory preserved, when regular power fails.

2. *Off-line programming* or off line compiling. Most process control computers do not have capacity for compiling their own programs *on-line*. The programs are compiled on a larger computer, put on tape, and played into the memory of the process computer. In very large supervisory systems with back-up computers the off-line computer may sometimes be used for compilation.

3. *Boot-strapping* refers to the process of starting-up a computer, inputting its program material, and *educating* it for the tasks to be performed.

10.5   THE ECONOMICS OF SUPERVISORY SYSTEMS

Almost the entire justification for supervisory systems is the potential savings in energy and personnel. Originally, labor-saving was more important, especially in large campus-type installations where the ability to monitor and trouble-shoot from one central location saves thousands of man-hours annually. The advent of the *energy crisis* made *energy conservation* (a better term than *energy saving*) the more important factor. The so-called *optimization* programs were developed for the express purpose of maximizing efficiency and minimizing energy consumption.

The potential savings in any given situation may be estimated by comparing unsupervised operation to a carefully supervised and controlled operation. The savings must then be compared to the amortization cost of the supervisory system for justification.

In a large campus-type situation with many buildings, the economics of the supervisory system can be readily demonstrated, especially when the system includes extensive special programs for optimization and maintenance. But what of smaller installations? It is this author's belief that some type of supervisory control can be justified for any commercial building. The simple systems described earlier in this chapter can be installed at very low cost, but even something as simple as a central hardwired panel with a time clock or two for automatic start-up and shut-down can save operating cost. The simpler special-purpose computer-type systems are useful in even fairly large installations.

The designer must therefore ask "What does the owner want, what does the owner need, and what can the owner afford?" The system that fits these criteria should be recommended and used.

10.6 BENEFITS OF THE COMPUTER SYSTEM

The computerized system has some interesting and valuable side benefits (for only slight additional costs).

10.6.1 Optimization

One of these side benefits is a program which will continuously analyze performance of the various systems, compare actual performance with some "ideal" standard and make adjustments to approach this standard and thus improve efficiency.

10.6.2 Data Management

Since the computer has already acquired a great mass of data about operating conditions, it is a simple matter to have this digested and printed out on any schedule and in any form desired. Some of that data may be very valuable for evaluating the existing design and improving future designs.

10.6.3 Maintenance Schedules

The regular preventive maintenance programs can be stored in the computer memory and printed out on schedule to remind the operator that they must be performed. As a further check, he must notify the computer of proper performance on schedule, or a follow-up notice will be printed by the computer.

Maintenance sensors can be made a part of the monitoring equipment. Such sensors will detect overheating of motors or bearings, excessive vibration, loud and extraneous noises and numerous other malfunction indicators.

10.6.4 Energy Consumption

Where a central chilled water or heating system serves a number of tenants, it may be necessary to meter the services supplied to each tenant. Special meters with outputs which can be read by the computer may provide an automatic measuring and billing service. Further, any sudden change from normal in a tenant's consumption would be instantly detected, and cause a warning message to be printed for investigation by operating personnel.

10.7 SUMMARY

We have discussed control systems for heating, ventilating and air conditioning systems, beginning with elements and ending with large supervisory systems. It is hoped that you have a better understanding of HVAC controls and, more importantly, control philosophy. Some of the ideas presented here may conflict with what you have learned previously. You may see what appear to be better ways of doing certain things. By all means, try something new if it looks possible. A great deal of the information presented here was learned "by mistake." Control design is interesting, challenging and fun. Enjoy it!

CONTROL BIBLIOGRAPHY

1. *Terminology for Automatic Control* (ASA C85,1–1963).
2. D. M. Considine, et al.: *Process Instruments & Controls Handbook* (McGraw Hill Book Co., New York, 1957).
3. American Society of Heating, Refrigeration and Air Conditioning Engineers: *Guide and Data Book, 1970, Systems,* Chapter 32, "Controls."
4. N. Peach: "Motor Control," *Power,* December 1962, pp S–1–S–20.
5. E. J. Brown: "How to Select Multiple Leaf Dampers for Proper Air Flow Control," *Heating Piping and Air Conditioning,* April 1960, pp. 167–168.
6. W. M. Marcinkowski: "How to Select Controls for End Use Exchangers in HTW Systems," *Heating Piping and Air Conditioning,* March 1964, pp 141–156.
7. "Understanding, Applying Solid State Devices," *Heating Piping and Air Conditioning,* November 1968, pp 115–162.
8. J. E. Haines: *Automatic Control of Heating and Air Conditioning* (McGraw Hill Book Co., New York, 1961, 2nd ed.).
9. R. C. Mott: "Building Blocks for Fluidic Control," *Heating Piping and Air Conditioning,* June 1968, pp 103–106.
10. R. E. Wagner: "Fluidics—A New Control Tool," *IEEE Spectrum,* November 1969, pp 58–59.
11. P. M. Kintner: "Interfacing a Control Computer with Control Devices," *Control Engineering,* November 1969, pp 97–101.
12. "Fluidic Output Interfaces," *Automation,* September 1969, pp 54–58.
13. Honeywell, Inc.: *Basic Electricity.*
14. Johnson Service Co.: *Field Training Handbook,* Fundamentals of Pneumatic Control.
15. Ibid., *Fundamentals of Electronic Control Equipment.*
16. Ibid., *Engineering Report No. 685,* "Electric Heat and Its Control."
17. Barber Coleman Co.: *Control Application Data,* manual Nos. CA–5–1, CA–18, CA–19–1, CA–24, CA–27, CA–28.
18. Ibid., *Automatic Controls Bulletins,* Nos. EN–54–1 and EN–61–1.
19. Ibid, *Pneumatic Controls Practices and Typical Devices* (F–13955).
20. National Environmental Systems Contractors Assn.: *Automatic Controls for Heating and Air Conditioning Systems.*
21. R. W. Haines: "The Economy Cycle, How Economical?" *Heating Piping and Air Conditioning,* June 1969, pp 86–88.
22. Roger W. Haines: "How to Pipe Recirculating Pumps for Preheat Coils," *Heating Piping and Air Conditioning,* July 1971, pp 56–60.

Abbreviations Used in This Book

A   Amperes
AC   Alternating current
Btu   British thermal units
C   Common
CB   Circuit breaker
CHR   Chilled water return
CHS   Chilled water supply
CHW   Chilled water
CWR   Condensing water return
CWS   Condensing water supply
DA   Direct-acting
DC   Direct current
DPDT   Double-pole, double-throw
DX   Direct-expansion
EP   Electric-pneumatic
ft   Feet
HOA   Hand-off-auto
hp   Horsepower
HTWR   High-temperature water return
HTWS   High-temperature water supply
HVAC   Heating, ventilating and air conditioning
HWR   Hot water return
HWS   Hot water supply
IC   Instantaneous closing
in.   Inches
IO   Instantaneous opening
KW   Kilowatt

MAX   Maximum
MIN   Minimum
min   minutes
NC   Normally closed
NO   Normally open
OL   Overload
PB   Pushbutton
PD   Pressure drop
PE   Pneumatic-electric
PRV   Pressure-reducing valve
PSIA   Pounds per square inch, absolute
PSIG   Pounds per square inch, gage
RA   Reverse-acting
RG   Refrigerant gas (discharge)
RL   Refrigerant liquid
RS   Refrigerant suction
SCFM   Standard cubic feet of air per minute
SCIM   Standard cubic inches of air per minute
sec   Seconds
SP   Static pressure
SPST   Single-pole, single-throw
SS   Start-stop
TC   Timed closing
TO   Timed opening
TR   Tower return
TS   Tower supply
V   Volts

Symbols Used in This Book

—| |— Capacitor

 Centrifugal fan

—\|— Check valve

CB
—o‿o— Circuit breaker

(V) Coil for solenoid valve

——▶ Contact, or point of force application

—\|⊦— Contact, NC

—| |— Contact, NO

[C] Contactor or motor starter; controller

(M) Control air supply

Control valve, three-way

Control valve, two-way

[DM] Damper motor

Diaphragm

[EP] Electric-pneumatic relay

(1M) Electromagnetic coil in starter or relay with identifier. All contacts actuated by this coil have the same identifier.

[FS] Fire safety switch

Float switch

Float valve

⬚ F   Flow switch

Flow switch, NC

Flow switch, NO

Fuse

Gas pilot flame with thermocouple

Globe valve

Ground connection

HAND
OFF
AUTO   Hand-off-auto (HOA) switch

⟋\/\/\⟍ Heater (heating element) or resistor

Ⓗ **Humidistat, room**

Industrial-type recording controller

Inlet vane damper (for centrifugal fan)

Limit switch

Logic relay

Manual switch

Manual switch with thermal overload

Motor

Motor field coil

Multipole switch (disconnect switch)

Needle valve

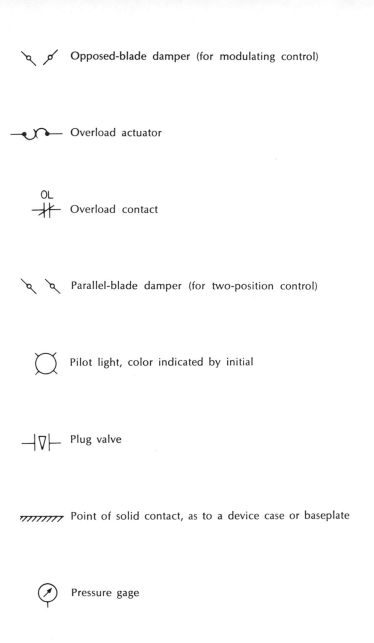

Opposed-blade damper (for modulating control)

Overload actuator

OL
Overload contact

Parallel-blade damper (for two-position control)

Pilot light, color indicated by initial

Plug valve

 Point of solid contact, as to a device case or baseplate

Pressure gage

Pressure regulator (pressure-reducing valve)

P Pressure switch or sensor

Pressure switch, NC

Pressure switch, NO

Propeller fan and motor

PC Proportioning controller, solid-state

Pump

 Pushbutton, normally closed (PB, NC)

Pushbutton, normally open (PB, NO)

R Relay

OR
 Relay coil

Relay or starter contact, NC

Relay or starter contact, NO

Relief valve

Resistor

Resistor, variable

RR   Reversing relay

SC   Sequencing controller

S   Smoke detector

Solenoid valve; solenoid valve, two-way

Solenoid valve, three-way

SPDT switch

SPDT switch with center-off position

 Spray nozzle

 Spring (where identified as such)

 SPST switch

 Static pressure controller

 Steam trap

 Thermal expansion valve, thermostatic expansion valve

Thermal switch, NC

Thermal switch, NO

Thermometer, remote-bulb or insertion type

T Thermostat, insertion type

T Thermostat, remote-bulb, duct or pipe, or insertion type

T Thermostat, room

Three-pipe control valve

IO Time-delay switch, NC, instantaneous open after energizing, timed
TC close after deenergizing

TO Time-delay switch, NC, timed open after energizing, instantaneous
close after deenergizing

IC Time-delay switch, NO, instantaneous close after energizing, timed
TO open after deenergizing

TC Time-delay switch, NO, timed close after energizing, instantaneous
open after deenergizing

Transformer

Transformer coil

L1 Wiring terminal with identification
o

229

Index